Metals, Fertility, and Reproductive Toxicity

Metals, Fertility, and Reproductive Toxicity

Edited by
Mari S. Golub

Taylor & Francis
Taylor & Francis Group
Boca Raton London New York

A CRC title, part of the Taylor & Francis imprint, a member of the
Taylor & Francis Group, the academic division of T&F Informa plc.

Published in 2006 by
CRC Press
Taylor & Francis Group
6000 Broken Sound Parkway NW, Suite 300
Boca Raton, FL 33487-2742

International Standard Book Number-10: 0-415-70040-X (Hardcover)
International Standard Book Number-13: 978-0-415-70040-5 (Hardcover)
Library of Congress Card Number 2005049128

Library of Congress Cataloging-in-Publication Data

Metals, fertility, and reproductive toxicity / edited by Mari Golub.
        p. cm.
    Includes bibliographical references.
    ISBN 0-415-70040-X (alk. paper)
    1. Metals--Toxicology. 2. Reproductive toxicology. I. Golub, Mari S.

RA1231.M52M53 2005
615.9'253--dc22   2005049128

Taylor & Francis Group
is the Academic Division of Informa plc.

Visit the Taylor & Francis Web site at
http://www.taylorandfrancis.com

and the CRC Press Web site at
http://www.crcpress.com

Editor

Mari S. Golub holds appointments as adjunct professor in the Department of Environmental Toxicology, University of California, Davis and staff toxicologist at the California Environmental Protection Agency. Her graduate training in psychopharmacology and toxicology were obtained from the University of Michigan and the University of California, Davis, with additional postdoctoral work at Boston University. She is certified by the American Board of Toxicology.

She has served on toxicology review groups for the National Institutes of Health, on expert panels for the National Toxicology Program's Center for Evaluation of Risk to Human Reproduction, and participated in a recent U.S. Environmental Protection Agency workshop to establish a Framework for Metals Risk Assessment.

Dr. Golub's research includes work on the developmental neurobehavioral toxicity of trace metal deficiency (zinc, iron) and excess (aluminum, manganese, arsenic). At the California Environmental Protection Agency, she contributes to the Proposition 65 program, which identifies agents as reproductive and developmental toxicants. Among the metals she has worked with for this program are arsenic, tin, chromium, and mercury.

Contributors

Michael J. Anderson
Office of Spill Prevention and
 Response
California Department of Fish and
 Game
Sacramento, California

Jose L. Domingo
Laboratory of Toxicology and
 Environmental Health
School of Medicine, "Rovira i Virgili"
 University
San Lorenzo, Reus, Spain

Makoto Ema
Division of Risk Assessment,
 Biological Safety Research Center
National Institute of Health Sciences
Tokyo, Japan

Mari S. Golub
Department of Environmental
 Toxicology
University of California
Davis, California

Akihido Hirose
Division of Risk Assessment,
 Biological Safety Research Center
National Institute of Health Sciences
Tokyo, Japan

Patricia B. Hoyer
Department of Physiology
University of Arizona
Tucson, Arizona

Wendie A. Robbins
UCLA Center for Occupational and
 Environmental Health
Los Angeles, California

Rebecca Z. Sokol
Department of Obstetrics and
 Gynecology and Medicine
Keck School of Medicine
University of Southern California
Los Angeles, California

Hilary Waites
Office of Spill Prevention and
 Response
California Department of Fish and
 Game
Sacramento, California

Julie T. Yamamoto
Office of Spill Prevention and
 Response
California Department of Fish and
 Game
Sacramento, California

Contents

chapter 1

Introduction

Mari S. Golub

Contents

Scope of This Volume

The topic of metals' reproductive toxicity is clearly too large to cover comprehensively in a single volume. There can be more than 30 elements designated as *metals*, depending on the definition used. Further, the number of metal agents is amplified manyfold when the various salts and organometals are considered. Issues related to reproductive health are also very broad and are relevant to many target populations, including wildlife, production (farm) animals, and pets, as well as human health. This volume focuses on human health data, and on laboratory animal data aimed at elucidating human health effects, adding a few important themes related to wildlife (lead and organotins). Metals with well-developed literatures in the reproductive toxicity area are also emphasized, along with more recent areas of interest such as *endocrine disruption*.

Other Sources of Information

A thorough review of the information generated on this topic prior to 1982 can be found in the 1983 volume *Reproductive and Developmental Toxicity of Metals*.* Reproductive toxicity is also one of the areas covered in the comprehensive chapter on metal toxicity provided by Goyer and Clarkson in the classic toxicology text *Cassaret and Doull's Toxicology: The Basic Science of Poisons*.**

---

* Clarkson, T.W., Nordberg, G.F., and Sager, P.R.. *Reproductive and Developmental Toxicity of Metals*, New York, Plenum Press, 1983.
** Klaassen, C.D., Ed. *Casarett & Doull's Toxicology: The Basic Science of Poisons*, 6th ed., New York, McGraw Hill, 2001.

Table 1.1 Toxicology Document Series Containing Information
on the Reproductive Toxicity of Individual Metals

	Current volume	ATSDR[a]	IPCS[b]	HC[c]	OEHHA[d]	USGS[e]
Aluminum	x	x	x		PHG	
Antimony		x				
Arsenic	x	x	x	x	HID,REL	x
Barium		x	x			
Beryllium		x			DS	
Boron		x	x		DS	x
Cadmium	x	x	x	x	HID, PHG	x
Chromium	x	x	x	x	PHG, DS	x
Cobalt		x				
Copper		x	x		PHG, REL, DS	x
Lead	x	x	x		PHG	x
Lithium						
Manganese		x	x		REL, DS	
Mercury	x	x	x		PHG, REL	x
Molybdenum						
Nickel		x	x	x	PHG, REL	x
Osmium						
Platinum			x			
Radium		x				
Selenium		x	x			x
Silver		x				x
Thallium		x	x		PHG	
Thorium		x				
Titanium		x	x			
Tin	x	X	x	x		x
Uranium	x	x			PHG	
Vanadium	x	x	x		REL	
Zinc		x	x			x

[a] Agency for Toxic Substances and Disease Control. Centers for Disease Control. Toxicological Profiles, http://www.atsdr.cdc.gov/toxpro2.html

[b] International Programme on Chemical Safety: World Health Organization. Environmental Health Criteria, http://www.who.int/pcs/pubs/pub_ehc.alph.htm

[c] Health Canada: Existing Substances Division. Priority Substances List, http://www.hc-sc.gc.ca/hecs-sesc/exsd/psap.htm

[d] Office of Environmental Health Hazard Assessment, http://www.oehha.ca.gov
Health risk assessment documents from the OEHHA include those directed exclusively at reproductive toxicity (Hazard Identification Documents [HIDs] and Data Summaries [DS]) as well as more general toxicity reviews that include reproductive toxicity (Public Health Goals [PHG] for drinking water, Reference Exposure Levels [REL] for air contaminants).

[e] Patuxent Health Center: United States Geological Survey. Contaminants Hazard Review, http://www.pwrc.usgs.gov/contaminants

Further information on the reproductive toxicity of individual metals can be found in regulatory documents, which generally review the toxicity of these agents (Table 1.1). Information on online access to these document series is provided in the footnotes.

Other online data sources specifically directed at reproductive toxicity include the DART database of the National Library of Medicine (http://toxnet.nlm.nih.gov/cgi-bin/sis/htmlgen?DARTETIC) and the Reprorisk® database of Micromedex (subscription required) (http://www.micromedex.com/products/reprorisk/). The ECOTOX database of the U.S. Environmental Protection Agency (USEPS, http://www.epa.gov/ecotox) allows searches using specific metals and keywords related to reproduction to locate resources relevant to wildlife populations. Many pesticides contain metals as active ingredients, and the required reproductive and developmental toxicity testing of these pesticides is reviewed in Registration Eligibility Decision documents from the USEPA (http://cfpub.epa.gov/oppref/reref/status.cfm?show=rereg).

A large amount of data concerning metal reproductive toxicity has been generated, summarized, and reviewed over the years. However, elucidation of integrating principles and characteristic actions at the cellular and molecular level that are relevant to reproductive harm await the work of new generations of reproductive toxicologists. The authors of this book hope to contribute to the advancement of this understanding through the reviews presented here.

chapter 2

Reproductive Toxicity of Mercury, Cadmium, and Arsenic

Mari S. Golub
Department of Environmental Toxicology
University of California, Davis

Contents

Introduction

Mercury and cadmium are well-known heavy metal toxicants with a broad spectrum of target organ toxicity, including reproductive toxicity. A number of reviews of the target organ toxicity of mercury and cadmium can be found in the peer-reviewed and regulatory agency literature that include, but do not focus on, fertility and reproductive harm (see Chapter 1). In addition, there is a newer set of reports concerning the estrogenic effects of mercury, cadmium, and arsenic relevant to the *endocrine disruption* theme. Ecotoxicological information is also growing for each of these agents, as they are more routinely monitored and studied in wildlife.

Arsenic is a metalloid whose reproductive toxicity has been primarily studied in connection with teratogenic effects in animal models. Recently, human studies of pregnancy outcome have become available, along with animal studies of effects on steroid hormone production at lower doses.

This chapter includes summaries from reviews and secondary sources, along with more detailed descriptions of recent and key studies.

Mercury

Overview

Mercury has been studied primarily for its neurotoxic effects in humans and animals. The largest developmental toxicity database is for developmental neurotoxicity, usually assessed postnatally, but data have been gathered on pregnancy outcome parameters in laboratory animals. In the reproductive toxicity area, occupational exposures have been of concern. The use of mercury in dental amalgams has led to study of reproductive health of dentists and dental assistants. There is a fairly large and long-standing literature on the effects of mercury on testes and sperm in laboratory animals. Finally, experimental studies of mercury in birds and fish demonstrate hormonal effects relevant to endocrine disruption.

Developmental Toxicity: Pregnancy Outcome

The neurodevelopmental toxicity of methyl mercury has been widely studied in both humans and animal models, with ongoing studies in several large human cohorts (NRC [National Research Council], 2000) using primarily functional behavioral endpoints. Because of the striking and sensitive effect of mercury on brain development, other pregnancy outcome endpoints have been less studied. However, reviews provided by government agencies (ATSDR [Agency for Toxic Substances and Disease Registry], 1994; USEPA [United States Environmental Protection Agency], 1997; and the NRC [National Research Council], 2000) describe effects after methyl mercury administration during pregnancy to rats and mice. At the higher doses, fetal viability was the major endpoint affected, followed by malformations (primarily craniofacial and central nervous system [CNS]) at midrange, and skeletal ossification and variation effects at lower doses, along with edema and kidney effects. Inorganic mercury exposure during pregnancy has also been studied in rats and mice, although to a lesser extent. Fetal viability and growth retardation were commonly affected in these studies, which often used injection routes, but also included gavage administration. Mercury accumulation in the placenta and possible interference with placental function has been studied as a mechanism of fetal effects (USEPA, 1994).

Female Fertility

There has been very little research on the effects of mercury in environmental media (air, water, soil) on human fertility. Of greatest concern has been the reproductive health of those exposed in the workplace, in particular the exposure of dentists and dental assistants because of the use of mercury (inorganic) in dental amalgams. Women dentists were found to have comparable fertility, in terms of time-to-pregnancy, as female schoolteachers (Dahl et al., 1999). In later studies of female dental assistants working with amalgam materials, fertility (measured as fecundability) was found to be lower in a highly exposed group (based on a questionnaire) than in a control group not working with these materials (Rowland et al., 1994). However at low mercury exposure levels, fertility was improved over controls. Similar enhanced fertility at low dose mercury exposure has been reported in quail and fish (see below).

Work in animal models concerning female fertility is also limited. Inorganic mercury exposure by inhalation of vapor (0, 1, 2 mg Hg/m^3) for 6 days before mating did not influence fertility of female rats (Davis et al., 2001). A similar 6-day exposure after mating did not affect implantation. At 2 mg/m^3, estrous cycle was lengthened. In a study exposing female monkeys to methyl mercury, conception was less frequent, and abortion and stillbirth more frequent than in controls (Burbacher et al., 1987–88). Mercury effects on the ovary are reviewed in Chapter 7.

Mercury has been known as an abortifacient in folk medicine. Using a similar paradigm (acute administration at high doses after implantation), hamsters were given a single oral dose of inorganic mercury (mercuric acetate) on gestation day (gd) 8 (Gale, 1974). Resorptions increased in a dose-response pattern at doses of 22, 32, 47, and 63 mg/kg/day. Methyl mercury had a similar effect on resorption in mice after gd 8 administration (Fuyuta et al., 1978) and in guinea pigs at several postimplantation times (Inouye and Kajiwara, 1988).

Endocrine Disruption in Wildlife

Reproductive failure is one of many pathologies noted in mercury-poisoned wildlife populations. A small number of experimental studies in birds and fish document low-level exposure effects on the endocrine axis. A series of doses of inorganic mercury (mercuric chloride) were administered in feed to quail after hatching and through sexual maturity. Sexual maturity (egg laying) appeared earlier in the treated hens, and greater fertility (more eggs laid) was noted. However, egg fertilization was reduced. There was no effect on hatchability (Hill and Shaffner, 1976).

An experimental study in fish demonstrating effects on gonadal hormones (Drevnick and Sandheinrich, 2003) has also become available recently using the fat head minnow model common in toxicology studies. Methylmercury in the diet beginning prior to sexual maturation led to decreased testosterone in males and increased estrogen in females after sexual maturity had been attained. Two concentrations of methyl mercury were provided in the feed, both of which increased mercury tissue concentrations, and were said to be within the range of exposures from contaminated waters. Reproductive organ growth and reproductive success were also adversely influenced, although growth rates were normal. Interestingly, fecundity appeared to be enhanced in the low dose exposure group, similar to the results of the quail studies and studies of dental assistants.

Male Reproductive Toxicity

Early studies documented adverse effects of inorganic mercury and methyl mercury on male fertility after intraperitoneal (i.p.) injection. In a serial mating paradigm, fertility was reduced beginning 20 days after a single 1 mg Hg/kg injection of either mercury chloride or methylmercury (Lee and Dixon, 1975). Reduced sperm populations were also seen, along with a *dominant lethal* effect (fewer live offspring). Dominant lethal effects were also reported in another mouse study (Suter, 1975). Effects on testes and spermatogenesis were later recorded with organic mercury administration in feed (Mitsumori et al., 1990). Studies in monkeys using 25 or 35 µg Hg/kg/d (gavage, 20 weeks) found effects on sperm motility and morphology (Mohamed et al., 1986; Mohamed et al., 1987) without significant depression of testosterone or histopathological abnormalities in testes biopsies.

Rao and colleagues have published a series of studies of mercury effects on testes in mice using oral administration (Rao, 1989a; Rao, 1989b; Rao and Sharma, 2001). In a fairly comprehensive study, inorganic mercury was administered by gavage to adult mice throughout one sperm cycle (45 days). Mice were examined at the end of treatment or after a 45-day recovery period. The dose used (1.25 mg $HgCl_2$/kg/d) did not influence weight or weight gain, but led to a 20 to 30% decrease in testes and cauda epididymides weight at the end of treatment. Mercury concentrations were two- to three-fold higher in testes and accessory organs at the end of treatment. Sperm from the cauda were reduced in number, viability, and motility, and no successful mating (sperm in vaginal lavage) was seen in a mating trial with 10 mice/group. Biochemical analysis of testes and accessory organs indicated altered antioxidant and steroid metabolizing activity; circulating testosterone was reduced. After the 45-day recovery period, most parameters, and in particular sperm parameters, were still depressed, although half the mice were able to mate.

As regards human exposures, no thorough epidemiological studies are available. Several studies of male workers and patients in infertility clinics have found potential associations between mercury exposure and sperm parameters (Schuurs, 1999). In addition, there was a suggested elevation in spontaneous abortion in wives of men exposed to mercury in an industrial setting in one study (Cordier et al., 1991), but not in another (Alcser et al., 1989). Male dentists have not been well studied, but existing data did not support an association between mercury exposure and spontaneous abortion (Brodsky et al., 1985).

Cadmium

A 1996 review (OEHHA [Office of Environmental Health Hazard Assessment], 1996) chronicled a large literature on cadmium reproductive toxicity, including animal studies with pregnancy administration, fertility mating trials in animals with both males and females exposed, and human studies correlating cadmium exposure or biomarkers (blood, hair, urine cadmium) and reproductive system impairment.

Chapter 7 contains a review of the effects of cadmium on the ovary and ovarian hormone production. The present review focuses on cadmium effects on birth weight, a major theme in human epidemiological studies, and recent work on the estrogenicity of cadmium.

Birth Weight Studies

Birth weight is the most extensively studied adverse pregnancy outcome associated with cadmium exposure in humans, and placental cadmium accumulation has been discussed as the possible mechanism. This issue initially arose in investigating the agent in cigarette smoke responsible for low birth weight in babies of smokers after cadmium accumulation was documented in the blood and placentas of pregnant smokers (Kuhnert et al., 1982). Kuhnert et al. (Kuhnert et al., 1987a; Kuhnert et al., 1987b) found a correlation of blood and placental cadmium with birth weight in a group of 77 smokers in Cleveland, Ohio, suggesting that cadmium was the agent responsible for intrauterine growth retardation associated with cigarette smoking. A second study found a supportive association between placental cadmium and birth weight in a population including smokers (n = 25) (Ward et al., 1987); however a study of 100 smokers in Poland failed to find an association between blood cadmium at birth and birth weight (Sikorski et al., 1988), although the effect of smoking was seen.

In women exposed to cadmium through the workplace or by living near smelters, no association of exposure with birth weight has been found (Berlin et al., 1992; Loiacono et al., 1992). Placental cadmium was the index of exposure in both studies, and coexposure to lead was also present in the Loicaono et al. study. However, an association between exposure and reduced birth weight has been investigated in a region of France with cadmium contamination from local industry (Huel et al., 1981; Huel et al., 1984; Bonithon-Kopp et al., 1986; Frery et al., 1993). In this area, infant hair cadmium concentrations were associated with low birth weight, but this association was modified by the presence of placental calcification. Placentae with calcification, an index of aging and degeneration of the placenta, had higher placental cadmium than those with no calcification. In pregnancies with calcified placenta (n=28), birth weight was negatively associated with infant hair cadmium, but positively associated with placenta cadmium. These associations suggest that cadmium accumulation and toxicity to the placenta leads to low birth weight and at the same time reduced cadmium transfer to the fetus.

More recently, a study of pregnancy outcome was conducted in an area of Japan known for prior episodes of cadmium poisoning (*itai-itai* disease) due to environmental contamination (Nishijo et al., 2002). Fifty-seven women were divided into low- and high-exposure groups based on urinary cadmium concentrations associated with clinical syndromes in other populations. The high exposure group (n = 12) differed from the low exposure group (n = 45) on a number of pregnancy outcome variables including pregnancy length, newborn height and weight, the number of low birth weight infants, and delivery by cesarean section. The authors felt that the effects on birth weight were mediated by early delivery, although they did not perform a "small for gestational age" analysis.

Drawing a conclusion across these various studies of cadmium exposure and birth weight is difficult because the sample used for the cadmium exposure marker (maternal blood, placenta, cord blood, infant hair, breast milk) varies, as well as the statistical methods used to test the association and the populations in which the association was investigated.

Lower birth weight was also a significant finding in a number of studies that administered cadmium (usually as cadmium chloride) to rats and mice by injection, inhalation, or oral (food and drinking water) routes (OEHHA, 1996). Delayed ossification, another indicator of developmental delay, was seen in many of these studies. At higher doses, fetal viability was affected as well.

One proposed mechanism of cadmium effects on birth weight is induction of metallothionein in the placenta and subsequent chelation of essential trace elements (zinc and copper). Zinc deficiency is known to cause malformations and intrauterine growth retardation in animal models. The interaction of zinc, copper, and cadmium in the placenta has been studied (Sowa et al., 1982; Steibert et al., 1984; Sowa and Steibert, 1985; Sasser et al., 1985; Kuhnert et al., 1987b; Kuhnert et al., 1993; Chmielnicka and Sowa, 1996). However, various supplement studies in animals including zinc supplementation, have not achieved reversal of cadmium effects on birth weight. Further, it has been suggested that metallothionein induction could enhance, rather than obstruct, zinc and copper transfer to the fetus (Goyer and Cherian, 1992). Current developments in identification and characterization of metal binding and metal transport proteins (Goyer and Clarkson, 2001; Ballatori, 2002) could help clarify cadmium/trace element relationships in the placenta and their relevance to fetal growth.

Cadmium Estrogenic Actions

Cadmium, copper, and zinc have also been suggested to act as steroid hormone receptors based on a very early paper (Young et al., 1977). Using a standard competitive binding approach, the investigators showed that zinc, cadmium, and copper reduced estradiol binding to human endometrium to 37, 35, and 28% of control at a 1 mM concentration. The lowest effective concentration for zinc was 50 micromolar; micromolar concentrations of copper were not studied. Magnesium, calcium, and barium had similar limited effects (87 to 93% control) on binding; manganese and lead (83 and 67% of control) effects were similar and somewhat greater. The purpose of this work was to explore the mechanism of action of the copper intrauterine device (IUD) in preventing implantation. However, this mechanism has never been fully elucidated. Nonetheless, it is probably fair to say that copper is the metal that has had the most impact on human fertility, albeit in therapeutic rather than toxicologic context.

More recently, cadmium was found to mimic the effects of estrogen on gene transcription using a reporter gene, as well as genes normally regulated by estrogen including the ER receptor. The mediation by estrogen receptor, and specifically ERα, was demonstrated in this system (MFC-7 cells) (Garcia-Morales et al., 1994). Further studies with transfected cells, chimeric receptors, and competitive receptor binding, demonstrated cadmium binding to the hormone-binding domain of ER. Cadmium binding could be weakened by targeted replacement of cysteine, glutamic acid, aspartic acid, and histidine in the hormone-binding region. Thus previously recognized interactions of cadmium with sulfur-containing amino acids could play a role.

In order to demonstrate that the *in vivo* effects of cadmium on the reproductive tract could be mediated through the estrogen receptor, an experiment was conducted in rats with cadmium (a single i.p. injection of cadmium chloride at 5 µg/kg), an estrogen receptor antagonist, and estradiol as a positive control for estrogenic action (Johnson et al., 2003). Uterotrophic and mammary gland development effects of cadmium were found to resemble those of estradiol and to be blocked by antiestrogen. Also, cadmium administered i.p. on gd 12 and 17 to rats led to altered pubertal markers in the female offspring. This cadmium effect on puberty was not compared to estradiol or examined for antagonism by antiestrogen, but was considered by the authors to be an estrogenic effect. As regards fertility, an early study using a single cadmium injection prior to mating in female mice increased corpora lutea at necropsy on gd 12 to 15 (Suter, 1975). A single cadmium injection (2.5, 5, or 10 mg/kg) on the day after mating failed to interfere with ova transport in the fallopian tubes (Paksy et al., 1992), although progesterone production was affected. Other studies concerning cadmium effects on ovarian hormone production are reviewed in Chapter 7.

The findings on cadmium estrogenicity have been confirmed in a different laboratory. In connection with screening for endocrine disruptors, a Japanese group conducted a screening of 20 metals using two screening assays (1) a transcription assay using a transformed breast cancer cell line, and (2) a proliferation assay in the same cell line (MFC-7) (Choe et al., 2003). More than one preparation of some metals was used, including some organic forms (tributyl tin, dimethyl mercury). Six metals showed substantial estrogenicity in both assays: tributyltin, cadmium, antimony, lithium, barium, and chromium. The potency of tributyltin was closest to that of betaestradiol (within an order of magnitude), while that of cadmium was about one order of magnitude that of estradiol. Chromium was effective as the chloride, but not as potassium chromate, while lithium was effective as the hydroxide but not the chloride. Four metals showing little or no activity in either assay were arsenic, cobalt, copper, and lead. Metals showing minimal or inconsistent effects across the two tests were mercury (dimethyl), lead (acetate), lithium chloride, magnesium, manganese (chloride and also methylcyclopentadienyl manganese tricarbonyl, MMT), tellurium, molybdenum, selenium, stannous chloride, silver titanium tungsten, and zinc. The authors did not provide any speculation as to why some metals were effective, but cited previous work implicating cadmium, lithium, barium, and chromium as reproductive toxicants. Neither valence nor atomic weight seemed relevant. Other divalent metals, such as zinc, which would be expected to interact with similar amino acid residues in the estrogen receptor, were not effective. Except for tributyltin, cadmium was the most estrogenic of the series of metals.

Arsenic

Arsenic as a Teratogen in Animal Studies

After the discovery of malformation induced by arsenic injection in the early 1970s, the teratogenic effects of arsenic continued to be explored in animal models. Several reviews are available (Willhite and Ferm, 1984; Shalat et al., 1996; DeSesso et al., 1998; Golub et al., 1998) describing studies that compared routes of administration (in feed, drinking water, by inhalation), forms of arsenic (arsenate, arsenite salts, and organic arsenic), and animal models (rats, mice, hamsters). In studying arsenic-induced teratogenesis, administration was usually during the period of organogenesis to maximize induction of malformations without general toxicity. The characteristic syndromes vary with dose and species, but primarily include neural tube closure defects, anophthalmia and microophthalmia, renal and gonadal agenesis, fused ribs, and vertebral and sternebral abnormalities. These studies clarified that arsenite was the most teratogenic form of arsenic; however, arsenate is more frequently used in these studies because arsenite also produces severe maternal toxicity, which can interfere with the detection of developmental effects. Mice and hamsters were more sensitive than rats, which may be related to differences in arsenic metabolism in these species (Vahter, 1999). Oral administration (gavage) was not as effective as injection on a mg/kg basis, but did produce characteristic malformations at low rates. A limited number of inhalation studies (relevant to arsenic exposure from smelter operations) include a minimally reported study in mice, a sensitive species, and a more recent and fully reported study in rats, an insensitive species for arsenic developmental toxicity. Since teratogenic doses of arsenic are also toxic to the mother, the role of maternal toxicity in secondarily influencing the fetus has been examined (Golub, 1994; Hood, 1998). Organic arsenic (primarily cacodylic acid) has been shown to produce a developmental toxicity syndrome similar to that of inorganic arsenic, although in a higher dose range (Hood, 1989).

Although the primary focus of the animal studies was malformation, other pregnancy outcome parameters affected included growth retardation and fetal death. Transplacental carcinogenesis and postnatal behavioral studies have more recently received some attention (Rodriguez et al., 2002; Waalkes et al., 2003). The transplacental cancer effect occurred in reproductive organs at environmentally relevant arsenic doses and may thus prove to be the most sensitive form of developmental toxicity.

The mechanism of arsenic-induced malformation was discussed early in terms of cell death, as detected in histological sections. More recently, an interference with methylation processes has been targeted for mechanism investigation (Wlodarczyk et al., 2001; Crandall and Vorce, 2002). This is a particularly interesting possibility because of the demonstrated efficacy of folate in preventing neural tube defects and the detoxification of arsenic via methylation. Actions mediated through transcription/growth factors have also been investigated. Mice demonstrating disruption of the PAX3 gene (splotch mutant), which is involved in neural tube closure, were more sensitive to arsenic-induced neural tube defects than wildtype mice of the same strain (Machado et al., 1999). However, the arsenic-induced incidence of skeletal malformations was also higher in the splotch mutants. Nonetheless, the authors suggest that arsenic may influence the PAX3 patterning pathways.

Pregnancy Outcome in Human Populations Exposed via Drinking Water

No studies have linked arsenic exposure and malformations in human populations. Arsenic teratogenesis would probably be most relevant to poisoning cases, rather than low-level chronic exposures that occur in human populations through environmental contamination. In particular, elevated drinking water arsenic is a major public health concern. In the past decade, the intensive study of human populations in Bangladesh, Chile, and China exposed to elevated arsenic in drinking water have greatly increased the amount of human data available concerning arsenic toxicity. Originally these studies were oriented toward cancer, but reproductive endpoints, primarily pregnancy outcome, have been reported more recently.

In a cross-sectional study (Ahmad et al., 2001), the incidences of three adverse pregnancy outcomes (spontaneous abortion, stillbirth, preterm birth) were determined by interviews using a questionnaire and checklist. Respondents (n = 96 per group) were randomly selected from the exposed population, and controls were matched for age, education, and socioeconomic status. Arsenic concentrations in drinking water were obtained from a national database. Statistical comparisons were made between a low-exposure community (drinking water concentration less than .02 mg As/L) and a high-exposure community (drinking water concentration greater than .05 mg As/L in 85% of wells). Subgroups within the high exposure community with briefer and longer exposures (5 to 15 years or 15 years or more) were also compared. Comparisons were significant for all three endpoints between exposed and nonexposed groups, as well as between shorter and longer exposure groups. Arsenic concentrations in drinking water were obtained from a national database. The average arsenic concentration of the wells in the exposed community was 0.240 mg/L.

Three supporting studies also found associations between arsenic drinking water exposure and the endpoints from the Ahmad et al. study (spontaneous abortion, stillbirth, and preterm birth). The first study was a brief report in which spontaneous abortion and stillbirth were examined in a Hungarian population exposed to elevated arsenic in drinking water from regional wells (Borzsonyi et al., 1992). The arsenic-exposed population (n > 5000 pregnancies) was compared to a population in a neighboring area with low drinking water arsenic concentrations (n > 2000 pregnancies). The arsenic-exposed population had significantly elevated rates of spontaneous abortion and stillbirth. The second study (Yang et al., 2003) found a nonsignificant increase in preterm birth in a geographical area in Taiwan with elevated drinking water arsenic concentrations, compared to a nonexposed area. This study used only first-parity singleton live births and controlled for maternal age, marital status, maternal education, and sex in the regression. Yang et al. (2003) also found a significantly lower birth weight (an endpoint not considered by Ahmad et al.) in the arsenic-exposed population. The third study (Hopenhayn-Rich et al., 2000) compared pre- and postnatal mortality in two communities in Chile during a period of high exposure to arsenic from drinking water in one of the communities. Higher rates of stillbirth (defined as fetal death after 28 weeks of gestation) were significantly associated with the arsenic exposure. Hopenhayn-Rich et al. also found a significant association of arsenic exposure with elevated neonatal and infant death.

These three studies provide supportive evidence of an association of arsenic exposure with spontaneous abortion, stillbirth, and preterm birth. One study has been directed exclusively at birth weight (Hopenhayn et al., 2003). It used a prospective cohort design to look at two communities in Chile that have historically differed in their arsenic exposure via drinking water (40 µg/L vs. < 1 µg/L). Birth weights obtained from medical records for approximately 420 newborns in each location were entered into a multivariate regression analysis along with sex, gestation length, and parity, age, BMI, height, and prenatal care of the mothers. The arsenic-exposed group had birth weight 59 g lower than the unexposed group. However, the 95% confidence interval (–122 to +9 g) overlapped the value for the null hypothesis (0). The authors state that this size difference is in the range produced by environmental toxicant exposures such as environmental tobacco smoke (ETS) and benzene.

Animal Studies of Reproductive Toxicity with Drinking Water Exposure

These epidemiological studies have prompted the development of animal models with chronic administration in drinking water to investigate reproductive effects. Some progress in this area has been accomplished by research groups in South Asia, an area affected by epidemic arsenicosis.

The investigators used a drinking water concentration of 0.4 mg/L, in the midrange for contaminated drinking water in West Bengal, India, one of the known sites of drinking water arsenicosis, for a study of arsenic (arsenite) effects on female reproductive organs. Two durations of exposure, 16 and 28 days, were used with 10 rats per group. Based on drinking water consumption of 10 mL/day and body weight of 160 g, the dose was .025 mg/kg body weight per day. This paper (Chattopadhyay et al., 1999) found a reduction in reproductive organ weights (ovary 58%, uterus 76%, vagina 56%) after 28 days of exposure. Histologically, ovarian follicle regression and reduced thickness of the endometrium were seen (Chattopadhyay et al., 2003). Body weights were not affected and nonreproductive organs were not weighed. Steroid metabolizing enzymes of the ovary had about 50% reduced activity, and plasma estrogen levels were similarly reduced in the arsenic-exposed group. The effects at 0.4 mg/L in rats also included suppression of estrus cycling (permanent diestrous). Interestingly, plasma FSH and LH were also reduced. This suggested that the ovary was not being directly affected, but that gonadotropin release was suppressed by arsenic, subsequently reducing hormone production and maintenance of reproductive organ weights. The authors suggest that suppression of gonadotropin was secondary to enhanced ACTH production as demonstrated in earlier studies. Changes in neurotransmitter levels in midbrain support centrally regulated gonadotropin suppression. Further work by this research group has demonstrated that dietary supplementation with vitamin E, ascorbic acid, and selenium could counteract this effect of arsenic. These agents were chosen as antioxidants, but the potential role of oxidative damage in the syndrome has not been elucidated.

Arsenite in drinking water (40 mg/L for 35 days) was found to have effects in the testes and sperm of mice at higher doses than those that damaged ovaries of rats. Based on information in the article, this water concentration led to a dose of 286 mg As/kg/d. Lower concentrations (4, 10, 20 mg/L) were not effective. The authors stated that body weight was not affected. The steroid metabolizing enzyme 17 beta HSD was lower, and sperm count motility and morphology were affected (Pant et al., 2001). Similar effects have been demonstrated in rats, but using an injection route. Intraperitoneal arsenite at doses of 5 and 6 mg/kg for 26 days lowered testes, seminal vesicle, and ventral prostate weights in rats without a concomitant reduction in body weight (Sarkar et al., 2003). Arsenic concentrations in testes were elevated and plasma testosterone concentrations were reduced. Sperm count in the epididymis was reduced by 50%, and spermatid populations in the testes were reduced by 36%. The authors suggest a secondary effect on gonadotropins mediated by arsenic-induced adrenocortical activation, as previously proposed for ovarian effects (Chattopadhyay et al., 1999). However, the ovary appeared to be a much more sensitive target based on effective water concentration of 0.4 mg/L for ovarian damage and 40 mg/L for testicular damage.

Other arsenic-containing compounds have received some attention for testicular effects. As arsenic has come into therapeutic use for cancer chemotherapy (Chen et al., 2001), various arsenicals are receiving attention for possible male reproductive effects. In a structure-activity study using human sperm mobility as an endpoint (Uckun et al., 2002), organic arsenic compounds with high dipole moments were most effective when added to medium. Binding to thiol groups and oxidative damage were suggested as mechanisms of action. Previous studies failed to find a dominant lethal effect of arsenite (Golub et al., 1998). Gallium and indium arsenide, two arsenic compounds used in the semiconductor industry, were studied for testicular toxicity in rats and hamsters (Omura et al., 1996a; Omura et al., 1996b). Arsenic trioxide was included in the study, which used intratracheal instillation to simulate workplace inhalation exposures. Both agents caused testicular and sperm damage, but the pattern of effects across agents compared to arsenic blood levels did not suggest that arsenic was a common toxic agent.

No mating trials were conduced in connection with the recent studies of low-dose arsenic in drinking water and reproductive organ damage in rats (Chattopadhyay et al., 1999; Pant et al., 2001; Sarkar et al., 2003). These studies can be compared to earlier chronic dosing studies in which fertility was assessed. An early multigeneration study in mice (Schroeder and Mitchener, 1971) reported smaller litter sizes with drinking water concentrations of arsenite of 5 ppm (dose estimated at 1 mg/kg/d), considerably higher than the 0.4 ppm (0.025 mg/kg/d dose) drinking water concentration found to damage ovaries in rats. A more recent multigeneration study in mice reported smaller litter sizes using arsenate administered in diet at a dose of

13 mg/kg/d (Hazelton, Laboratories of America, 1990). Neither multigeneration study reported arsenic effects on fertility. The findings of smaller litter size in animal studies are potentially related to findings of spontaneous abortion and stillbirth in humans. However, it is not clear whether there is a connection between smaller litter size and ovarian effects of drinking water arsenic exposure. Parameters such as estrous cycling and puberty onset have not been studied in animals exposed to arsenic in drinking water.

Estrogenic Activity of Arsenic

An estrogenic effect of arsenic was proposed in connection with arsenate carcinogenesis studies conducted in mice (Waalkes et al., 2000). Although arsenic is a possible human carcinogen, it has not been found to induce tumors in laboratory rodents. In this study, arsenate was injected weekly in the tail vein for 20 weeks. Arsenic was not found to induce tumors, in agreement with earlier studies, but proliferative lesions of reproductive organs (uterus and testes) were identified in the pathology exam. Proliferative lesions were also seen in the dermis and liver. In support of an estrogenic action, the investigators reported upregulation of estrogen receptor in early stages of proliferation, as well as the estrogen-regulated protein cyclinD. The authors stated that arsenic action as an estrogen receptor activator was unlikely, and this is supported by later *in vitro* studies (Choe et al., 2003). Rather the authors suggested a modulation of estrogen action possibly through the nitric oxide pathway.

Further attention to arsenic reproductive and developmental toxicity, building on the older body of work and taking into account newer population-based human studies, may reveal a new and important aspect of the actions of this historic toxicant.

References

Ahmad, S.A., Sayed, M.H., Barua, S., Khan, M.H., Faruquee, M.H., Jalil, A., Hadi, S.A., and Talukder, H.K. 2001. Arsenic in drinking water and pregnancy outcomes. *Environ Health Perspect*, 109, 629–31.

Alcser, K.H., Brix, K.A., Fine, L.J., Kallenbach, L.R., and Wolfe, R.A. 1989. Occupational mercury exposure and male reproductive health. *Am J Ind Med*, 15, 517–29.

ATSDR (Agency for Toxic Substances and Disease Registry). 1994. *Toxicological profile for mercury (update)*. U.S. Department of Health and Human Services, Washington, DC.

Ballatori N. 2002. Transport of toxic metals by molecular mimicry. *Environ Health Perspect*, 110, Suppl 5, 689–94.

Berlin, M., Blanks, R., Catton, M., Kazantzis, G., Mottet, N.K., and Samiullah, Y. 1992. Birth weight of children and cadmium accumulation in placentas of female nickel-cadmium (long-life) battery workers. *IARC Sci Publ*, 257-62.

Bonithon-Kopp, C., Huel, G., Grasmick, C., Sarmini, H., and Moreau, T. 1986. Effects of pregnancy on the inter-individual variations in blood levels of lead, cadmium and mercury. *Biol Res Pregnancy Perinatol*, 7, 37–42.

Borzsonyi, M., Bereczky, A., Rudnai, P., Csanaday, M., and Horvath A. 1992. Epidemiological studies on human subjects exposed to arsenic in drinking water in southeast Hungary. *Arch Toxicol*, 66, 77–78.

Brodsky, J.B., Cohen, E.N., Whitcher, C., Brown, B.W., Jr, and Wu, M.L. 1985. Occupational exposure to mercury in dentistry and pregnancy outcome. *J Am Dent Assoc*, 111, 779–80.

Burbacher, T.M., Mohamed, M.K., and Mottett, N.K. 1987–1988. Methylmercury effects on reproduction and offspring size at birth. *Reprod Toxicol*, 1, 267–78.

Chattopadhyay, S., Ghosh, S., Chaki, S., Debnath, J., and Ghosh, D. 1999. Effect of sodium arsenite on plasma levels of gonadotrophins and ovarian steroidogenesis in mature albino rats: Duration-dependent response. *J Toxicol Sci*, 24, 425–31.

Chattopadhyay, S., Pal (Ghosh), S., Ghosh, D., and Debnath, J. 2003. Effect of dietary co-administration of sodium selenite on sodium arsenite induced ovarian and uterine disorders in mature albino rats. *Toxicol Sci*, 75, 412–422.

Chen, Z., Chen, G.Q., Shen, Z.X., Chen, S.J., and Wang, Z.Y. 2001. Treatment of acute promyelocytic leukemia with arsenic compounds: *In vitro* and *in vivo* studies. *Semin Hematol*, 38, 26–36.

Chmielnicka, J. and Sowa, B. 1996. Cadmium interaction with essential metals (Zn, Cu, Fe), metabolism metallothionein, and ceruloplasmin in pregnant rats and fetuses. *Ecotoxicol Environ Saf*, 35, 277–81.

Choe, S.Y., Kim, S.J., Kim, H.G., Lee, J.H., Choi, Y., Lee, H., and Kim, Y. 2003. Evaluation of estrogenicity of major heavy metals. *Sci Total Environ*, 312, 15–21.

Cordier, S. et al. 1991. Paternal exposure to mercury and spontaneous abortions. *Br J Ind Med*, 48, 375–81.

Crandall, L.Z. and Vorce, R.L.. 2002. Differential effects of arsenic on folate binding protein 2 (Folbp2) null and wild type fibroblasts. *Toxicol Lett*, 136, 43–54.

Dahl, J.E., Sundby, J., Hensten-Pettersen, A., and Jacobsen, N. 1999. Dental workplace exposure and effect on fertility. *Scand J Work Environ Health*, 25, 285–90.

Davis, B.J., Price, H.C., O'Connor, R.W., Fernando, R., Rowland, A.S., and Morgan, D.L. 2001. Mercury vapor and female reproductive toxicity. *Toxicol Sci*, 59, 291–6.

DeSesso, J.M., Jacobson, C.F., Scialli, A.R., Farr, C.H., and Holson, J.F. 1998. An assessment of the developmental toxicity of inorganic arsenic. *Reprod Toxicol*, 12, 385–433.

Drevnick, P.E. and Sandheinrich, M.B. 2003. Effects of dietary methylmercury on reproductive endocrinology of fathead minnows. *Environ Sci Technol*, 37, 4390–6.

Frery, N., Nessmann, C., Girard, F., Lafond, J., Moreau, T., Blot, P., Lellouch, J., and Huel, G. 1993. Environmental exposure to cadmium and human birthweight. *Toxicology*, 79, 109–18.

Fuyuta, M., Fujimoto, T., and Hirata S. 1978. Embryotoxic effects of methylmercuric chloride administered to mice and rats during orangogenesis. *Teratology*, 18, 353–66.

Gale, T.F. 1974. Embryopathic effects of different routes of administration of mercuric acetate in the hamster. *Environ Res*, 8, 207–13.

Garcia-Morales, P., Saceda, M., Kenney, N., Kim, N., Salomon, D.S., Gottardis, M.M., Solomon, H.B., Sholler, P.F., Jordan, V.C., and Martin, M.B. 1994. Effect of cadmium on estrogen receptor levels and estrogen-induced responses in human breast cancer cells. *J Biol Chem*, 269, 16896–901.

Golub, M. 1994. Maternal toxicity and the identification of inorganic arsenic as a developmental toxicant. *Reprod Toxicol*, 8, 283–295.

Golub, M.S., Macintosh, M.S., and Baumrind, N. 1998. Developmental and reproductive toxicity of inorganic arsenic: Animal studies and human concerns. *J Toxicol Environ Health B Crit Rev*, 1, 199–241.

Goyer, R.A. and Cherian, M.G. 1992. Role of metallothionein in human placenta and rats exposed to cadmium. *IARC Sci Publ* 239–47.

Goyer, R. and Clarkson, T. 2001. Toxic effects of metals, in Cassaret and Doull's Toxicology, Klaassen, C., Ed., McGraw-Hill, New York, 811–68.

Hazelton Laboratories of America. 1990. Two generation dietary reproduction study with arsenic acid in mice. Madison, Wisconsin.

Hill, E.F. and Shaffner, C.S. 1976. Sexual maturation and productivity of Japanese quail fed graded concentrations of mercuric chloride. *Poult Sci*, 55, 1449–59.

Hood R. 1989. A perspective on the significance of maternally mediated developmental toxicity. *Regul Toxicol Pharmacol*, 10, 144–48.

Hood, R.D. 1998. Developmental effects of methylated arsenic metabolites in mice. *Bull Environ Contam Toxicol*, 61, 231–8.

Hopenhayn, C., Ferreccio, C., Browning, S.R., Huang, B., Peralta, C., Gibb, H., and Hertz-Picciotto, I. 2003. Arsenic exposure from drinking water and birth weight. *Epidemiology*, 14, 593–602.

Hopenhayn-Rich, C., Browning, S.R., Hertz-Picciotto, I., Ferreccio, C., Peralta, C., and Gibb, H. 2000. Chronic arsenic exposure and risk of infant mortality in two areas of Chile. *Environ Health Perspect*, 108, 667–73.

Huel, G., Boudene, C., and Ibrahim, M.A. 1981. Cadmium and lead content of maternal and newborn hair: Relationship to parity, birth weight, and hypertension. *Arch Environ Health*, 36, 221–7.

Huel, G., Everson, R.B., and Menger, I. 1984. Increased hair cadmium in newborns of women occupationally exposed to heavy metals. *Environ Res*, 35, 115–21.

Inouye, M. and Kajiwara, Y. 1988. Developmental disturbances of the fetal brain in guinea-pigs caused by methylmercury. *Arch Toxicol*, 62, 15–21.

Johnson, M.D., Kenney N., Stoica, A., Hilakivi-Clarke, L., Singh, B., Chepko, G., Clarke, R., Sholler, P.F., Lirio, A.A., Foss, C., Reiter, R., Trock, B., Paik, S., and Martin, M.B. 2003. Cadmium mimics the *in vivo* effects of estrogen in the uterus and mammary gland. *Nat Med*, 9, 1081–4.

Kuhnert, B.R., Kuhnert, P.M., Debanne, S., and Williams, T.G. 1987a. The relationship between cadmium, zinc, and birth weight in pregnant women who smoke. *Am J Obstet Gynecol*, 157, 1247–51.

Kuhnert, B.R., Kuhnert, P.M., Lazebnik, N., and Erhard, P. 1993. The relationship between placental cadmium, zinc, and copper. *J Am Coll Nutr*, 12, 31–5.

Kuhnert, P.M., Kuhnert, B.R., Bottoms, S.F., and Erhard, P. 1982. Cadmium levels in maternal blood, fetal cord blood, and placental tissues of pregnant women who smoke. *Am J Obstet Gynecol*, 142, 1021–5.

Kuhnert, P.M., Kuhnert, B.R., Erhard, P., Brashear, W.T., Groh-Wargo, S.L., and Webster, S. 1987b. The effect of smoking on placental and fetal zinc status. *Am J Obstet Gynecol*, 157, 1241–6.

Lee, I.P. and Dixon, R.L. 1975. Effects of mercury on spermatogenesis studied by velocity sedimentation cell separation and serial mating. *J Pharmacol Exp Ther*, 194, 171–81.

Loiacono, N.J., Graziano, J.H., Kline, J.K., Popovac, D., Ahmedi, X., Gashi, E., Mehmeti, A., and Rajovic, B. 1992. Placental cadmium and birthweight in women living near a lead smelter. *Arch Environ Health*, 47, 250–5.

Machado, A.F., Hovland, D.N., Jr., Pilafas, S., and Collins, M.D. 1999. Teratogenic response to arsenite during neurulation: Relative sensitivities of C57BL/6J and SWV/Fnn mice and impact of the splotch allele. *Toxicol Sci*, 51, 98–107.

Mitsumori, K., Hirano, M., Ueda, H., Maita, K., and Shirasu, Y. 1990. Chronic toxicity and carcinogenicity of methylmercury chloride in B6C3F1 mice. *Fundam Appl Toxicol*, 14, 179–90.

Mohamed, M.K., Burbacher, T.M., and Mottet, N.K. 1987. Effects of methyl mercury on testicular functions in Macaca fascicularis monkeys. *Pharmacol Toxicol*, 60, 29–36.

Mohamed, M.K., Evans, T.C., Mottet, N.K., and Burbacher, T.M. 1986. Effects of methyl mercury on sperm oxygen consumption. *Acta Pharmacol Toxicol (Copenh)*, 58, 219–24.

National Research Council (NRC). 2000. *Toxicological effects of methyl mercury.* National Academies Press, Washington, DC.

Nishijo, M. Nakagawa, H. Honda, R., Tanebe, K., Saito, S., Teranishi H., and Tawara K. 2002. Effects of maternal exposure to cadmium on pregnancy outcome and breast milk. *Occup Environ Med*, 59, 394–6; discussion 397.

Office of Environmental Health Hazard Assessment (OEHHA). 1996. *Evidence on Reproductive and Developmental Toxicity of Cadmium.* California Environmental Protection Agency, Sacramento, CA.

Omura, M., Hirata, M., Tanaka, A., Zhao, M., Makita, Y., Inoue, N., Gotoh, K., and Ishinishi, N. 1996a. Testicular toxicity evaluation of arsenic-containing binary compound semiconductors, gallium arsenide and indium arsenide, in hamsters. *Toxicol Lett*, 89, 123–9.

Omura, M., Tanaka, A., Hirata, M., Zhao, M., Makita, Y., Inoue, N., Gotoh, K., and Ishinishi, N. 1996b. Testicular toxicity of gallium arsenide, indium arsenide, and arsenic oxide in rats by repetitive intratracheal instillation. *Fundam Appl Toxicol*, 32, 72–8.

Paksy, K., Varga, B., Naray, M., Olajos, F., and Folly, G. 1992. Altered ovarian progesterone secretion induced by cadmium fails to interfere with embryo transport in the oviduct of the rat. *Reprod Toxicol*, 6, 77–83.

Pant, N., Kumar, R., Murthy, R.C., and Srivastava, S.P. 2001. Male reproductive effect of arsenic in mice. *Biometals*, 14, 113–7.

Rao, M.V. 1989a. Effects of methylmercury on mouse epididymis and spermatozoa. *Biomed Biochim Acta*, 48, 577–82.

Rao, M.V. 1989b. Toxic effects of methylmercury on spermatozoa in vitro. *Experientia*, 45, 985–7.

Rao, M.V. and Sharma, P.S. 2001. Protective effect of vitamin E against mercuric chloride reproductive toxicity in male mice. *Reprod Toxicol*, 15, 705–12.

Rodriguez, V.M., Carrizales, L., Mendoza, M.S., Fajardo, O.R., and Giordano, M. 2002. Effects of sodium arsenite exposure on development and behavior in the rat. *Neurotoxicol Teratol*, 24, 743–50.

Rowland, A.S., Baird, D.D., Weinberg, CR., Shore, D.L., Shy, C.M., and Wilcox, A.J. 1994. The effect of occupational exposure to mercury vapour on the fertility of female dental assistants. *Occup Environ Med*, 51, 28–34.

Sarkar, M., Chaudhuri, G.R., Chattopadhyay, A., and Biswas, N.M. 2003. Effect of sodium arsenite on spermatogenesis, plasma gonadotrophins and testosterone in rats. *Asian J Androl*, 5, 27–31.

Sasser, L.B., Kelman, B.J., Levin, A.A., and Miller, R.K. 1985. The influence of maternal cadmium exposure or fetal cadmium injection on hepatic metallothionein concentrations in the fetal rat. *Toxicol Appl Pharmacol*, 80, 299–307.

Schroeder, H. and Mitchener, M. 1971. Toxic effects of trace elements on the reproduction of mice and rats. *Arch Environ Health*, 23, 102–6.

Schuurs, A.H. 1999. Reproductive toxicity of occupational mercury. A review of the
literature. *J Dent*, 27, 249–56.

Shalat, S., Walker, D., and Finnel, R. 1996. Role of arsenic as a reproductive toxin
with particular attention to neural tube defects. *J Toxicol Environ Health*, 48,
253–272.

Sikorski, R., Radomanski, T., Paszkowski, T., and Skoda J. 1988. Smoking during
pregnancy and the perinatal cadmium burden. *J Perinat Med*, 16, 225–31.

Sowa, B. and Steibert, E. 1985. Effect of oral cadmium administration to female rats
during pregnancy on zinc, copper, and iron content in placenta, foetal liver,
kidney, intestine, and brain. *Arch Toxicol*, 56, 256–62.

Sowa, B., Steibert, E., Gralewska, K., and Piekarski, M. 1982. Effect of oral cadmium
administration to female rats before and/or during pregnancy on the metallothionein level in the fetal liver. *Toxicol Lett*, 11, 233–6.

Steibert, E., Krol, B., Sowa, B., Gralewska, K., Kaminski, M., Kaminska, O., and Kusz,
E. 1984. Cadmium-induced changes in the histoenzymatic activity in liver,
kidney and duodenum of pregnant rats. *Toxicol Lett*, 20, 127–32.

Suter, K.E. 1975. Studies on the dominant-lethal and fertility effects of the heavy
metal compounds methylmercuric hydroxide, mercuric chloride, and cadmium chloride in male and female mice. *Mutat Res*, 30, 365–74.

Uckun, F.M., Liu, X.P., and D'Cruz, O.J. 2002. Human sperm immobilizing activity of
aminophenyl arsenic acid and its N-substituted quinazoline, pyrimidine, and
purine derivatives: Protective effect of glutathione. *Reprod Toxicol*, 16, 57–64.

United States Environmental Protection Agency (USEPA). 1994. *Summary review of health
effects associated with mercuric chloride: Health issue assessment.* Research Triangle
Park, U.S. Environmental Protection Agency, Washington, DC.

United States Environmental Protection Agency (USEPA). 1997. *Mercury Study Report
to Congress. Volume V: Health Effects of Mercury and Mercury Compounds.* United
States Environmental Protection Agency, Washington, DC.

Vahter, M. 1999. Methylation of inorganic arsenic in different mammalian species
and population groups. *Sci Prog*, 82 (Pt 1), 69–88.

Waalkes, M.P., Keefer, L.K., and Diwan, B.A. 2000. Induction of proliferative lesions
of the uterus, testes, and liver in Swiss mice given repeated injections of
sodium arsenate: Possible estrogenic mode of action. *Toxicol Appl Pharmacol*,
166, 24–35.

Waalkes, M.P., Ward, J.M., Liu, J., and Diwan, B.A. 2003. Transplacental carcinogenicity of inorganic arsenic in the drinking water: Induction of hepatic, ovarian,
pulmonary, and adrenal tumors in mice. *Toxicol Appl Pharmacol*, 186, 7–17.

Ward, N., Watson, R., and Bryce-Smith, C. 1987. Placental elemental levels in relation
to fetal development for obstetrically "normal" births: A study of 37 elements.
Evidence for effects of cadmium, lead and zinc on fetal growth, and for
smoking as a source of Cadmium. *Int J Biosocial Res*, 9, 63–81.

Willhite, C. and Ferm, V.H. 1984. Prenatal and developmental toxicology of arsenicals,
in, *Nutritional and Toxicological Aspects of Food Safety*, Friedman, M., Ed., Plenum Press, New York, 205–28.

Wlodarczyk, B., Spiegelstein, O., Gelineau-van Waes, J., Vorce, R.L., Lu, X., Le, C.X.,
and Finnell, R.H. 2001. Arsenic-induced congenital malformations in geneti-
cally susceptible folate binding protein-2 knockout mice. *Toxicol Appl Pharmacol*, 177, 238–46.

Yang, C.Y., Chang, C.C., Tsai, S.S., Chuang, H.Y., Ho, C.K., and Wu, T.N. 2003. Arsenic in drinking water and adverse pregnancy outcome in an arseniasis-endemic area in northeastern Taiwan. *Environ Res*, 91, 29–34.

Young, P.C., Cleary, R.E., and Ragan, W.D. 1977. Effect of metal ions on the binding of 17beta-estradiol to human endometrial cytosol. *Fertil Steril*, 28, 459–63.

chapter 3

Reproductive and Developmental Toxicity of Organotin Compounds

Makoto Ema and Akihiko Hirose
Division of Risk Assessment, Biological Safety Research Center,
National Institute of Health Sciences, Tokyo, Japan

Contents

Introduction

Organotin compounds are chemicals widely used in agriculture and industry (Piver 1973, World Health Organization 1980). Tetrasubstituted organotin compounds are mainly used as intermediates in the preparation of other organotin compounds. Trisubstituted organotin compounds have biocidal properties and are used in agriculture as fungicides and acaricides, as rodent repellents, and mulluscicides, and are widely used as antifoulants in ship paints and underwater coatings. Especially, triphenyltins (TPTs) and tributyltins (TBTs) have been used extensively in antifouling products such as algaecides and mulluscicides. Disubstituted organotin compounds are commercially the most important derivatives, and are mainly used in the plastics industry, particularly as heat and light stabilizers for polyvinyl chloride (PVC) plastics to prevent degradation of the polymer during melting and forming of the resin into its final products, as catalysts in the production of polyurethane foams, and as vulcanizing agents for silicone rubbers. Monosubstituted organotin compounds are used as stabilizers in PVC films. Widespread use of organotin compounds has caused increasing amounts to be released into the environment. The most important nonpesticidal route of entry for organotin compounds into the environment is through leaching of organotin-stabilized PVC in water (Quevauviller et al. 1991), and the use in antifouling agents, resulting in the introduction of organotin into the aquatic environment (Maguire 1991). Data are available regarding the detection of butyltin and phenyltin compounds in aquatic marine organisms (Sasaki et al. 1988, Fent and Hunn 1991, Lau 1991) and marine products (Suzuki et al. 1992, Belfroid et al. 2000, Tsuda et al. 1995, Ueno et al. 1999, Toyoda et al. 2000). Food chain bioamplification of butyltin in oysters (Waldock and Thain 1983), mud crabs (Evans and Laughlin 1984), marine mussels (Laughlin et al. 1986), Chinook salmon (Short and Thrower 1986), and dolphin, tuna, and shark (Kannan et al. 1996), and of phenyltin in carp (Tsuda et al. 1987) and horseshoe crab (Kannan et al. 1995) has been reported. These indicate that organotin compounds accumulate in the food chain and are bioconcentrated, and that humans can be exposed to organotin compounds via seafood. The World Health Organization (WHO) reported in 1980 that the estimated mean total daily intake of tin by humans ranged from 200 µg to 17 mg. Recently, Tsuda et al. (1995) reported that the daily intakes in Shiga prefecture in Japan were 0.7 to 5.4 µg in 1991 and 0.7 to 1.3 µg in 1992 for TPT and 4.7 to 6.9 µg in 1991 and 2.2 to 6.7 µg in 1992 for TBT. Toyoda et al. (2000) also showed that the daily intakes in Japanese consumers, based on analysis with the 1998 total diet samples, were 0.09 µg for TPT, 0 µg for diphenyltin (DPT), 1.7 µg for TBT, and 0.45 µg for dibutyltin (DBT). These values are lower than the acceptable daily intake for TPT according to the JMPR (Joint Meetings of the FAO [Food and Agriculture Organization] and World Health Organization

Panel of Experts on Pesticides Residues), 25 μg (World Health Organization 1992), and the guidance value for oral exposure to tributyltin oxide (TBTO), 18 μg (International Programme on Chemical Safety 1999a). Thus, the levels of organotin compounds in seafood are not considered to be sufficiently high to affect human health (Tsuda et al. 1995, Ueno et al. 1999). However, Belfroid et al. (2000) noted that more research on residual TBT levels in seafood is needed before a definitive conclusion on possible health risks can be drawn.

In recent years, adverse effects of environmental chemicals on the reproductive success of wildlife populations have been reported (Colborn et al. 1993). These phenomena may result from interference with the endocrine system. Disturbances of hormonal regulation during pre- and postnatal development may produce deleterious effects on reproduction and development. TPT and TBT are suspected to be endocrine disruptors (Japan Environment Agency 1998). TBT and TPT are known to have strong effects on the development of imposex (imposition of male sex characteristics on females) in the rock shell (Horiguchi et al. 1996, 1997a), and this condition may bring about reproductive failure and a consequent population decline.

Although the toxicity of organotins has been extensively reviewed (World Health Organization 1980, Snoeij et al. 1987, Winship 1988, Boyer 1989, International Programme on Chemical Safety 1999a, b), the reproductive and developmental toxicity of these compounds is not well understood. In this chapter, we summarize the findings of the studies on reproductive and developmental effects of organotin compounds.

Effects on Aquatic Organisms

Imposex on Gastropods

TBT causes reproductive toxic effects in marine gastropods, which were represented by some masculinizing effects including *imposex* or *pseudohermaphodi(ti)sm*. The imposition of male sex organs (a penis and vas deferens) on female mud snails (*Nassarius obsoletus*) was found in near harbors, and the degree of penis development and frequency of imposex were positively correlated to the seawater TBT concentration (Smith 1981a, b). Imposex has been induced experimentally by treatment with 4.5 to 5.5 μg/L of TBT compounds for 60 days. In field studies in southeastern England, imposex has been reported in declining populations of the common dogwhelk (Bryan et al. 1986, 1987, 1989, Gibbs and Bryan 1986, Davies et al. 1987, Gibbs et al. 1987).

Imposex has not just occurred at a regional level, but worldwide on a global scale. Imposex in dogwhelk was not only reported in England, but in Scotland, the Netherlands, and the coastline of the North Sea. Imposex in other whelk species occurred in Canada, West Africa, New Zealand, Australia, Malaysia, Singapole, Indonesia, and Japan (Fent 1996, Horiguchi et al. 1996). Imposex among prosobranchs is known to occur in around 70 species of 50 genera, although some species are less susceptible to TBT compounds (Fioroni et al. 1991, Fent 1996).

TPT also induced imposex in *Thais clavigera* at the same potency as TBT (Horiguchi et al. 1997a). Although, in *Nucwlla lapillus*, TPT did not induce imposex, tripropyltin (TPrT) had a small effect on the development of imposex (Bryan et al. 1988). DBT and monobutyltin (MBT) did not induce imposex in the gastropod species examined. Three trisubstitution compounds (TBT, TPT, TPrT) and monophenyltin (MPT) easily induced imposex in some species, among the eight organotins, i.e., MBT, DBT, TBT, tetrabutyltin (TeBT), MPT, DPT, TPT, and TPrT. (Bryan et al. 1988, Hawkins and Hutchinson 1990, Horiguchi et al. 1997b).

The early studies in the 1980s reached some common conclusions, which are described below (Eisler 2000). Imposex correlated with the body burden of tributyl- and dibutyltin, but not with the tissue concentration of arsenic, cadmium, copper, lead, silver, or zinc. Forty-one percent of females had male characteristics, when the body burden reached to 1.65 mg Sn/kg of dry soft parts, by exposing with 0.02 μg Sn/L for 120 days. Imposex in immature females is caused above the concentration of around 1 ng/L (Sn) in seawater. At higher concentrations of TBT, the oviduct had been blocked, resulting in sterilization. Declining dogwhelk populations could be caused by aborting capsules, sterility, and premature death, which were characterized by a moderate to high degree of imposex, fewer female functions, fewer juveniles, and scarcity of laid egg capsules.

There is also a great variety of gradations of imposex in different species. The intensity is characterized by a classification system, which distinguishes six stages with a few different types, mainly based on a Vas Deferens Sequence (VDS) index (Oehlmann et al. 1991). Imposex development occurred in three variations: (1) a small penis without penis duct, (2) a short distal vas deferens section, or (3) a short proximal vas deferens section (stage 1). At stages 2 and 3 the male sex characteristics of each type are developed continuously. Stage 4 is characterized by a penis with penis duct and a complete vas deferens, and represents the last stage of fertility. The reproductive failure or sterility is induced in later stages. At stage 5 the vagina is replaced with a small prostate gland, the vagina opening is blocked by vas deferens tissue, or the incompletion of the pallial oviduct closure occurs. Abortive egg capsules fill the lumen and vestibulum of the capsule gland and evoke an intense swelling of the gland at stage 6 (Bettin et al. 1996). High TBT exposure in the early stages of life induced gametogenesis or sex changes characterized by a suppression of oogenesis and commencement of spermatogenesis in females (Gibbs et al. 1988, Fioroni et al. 1991, Oehlmann et al. 1991, 1996, Horiguchi et al. 2002). It was thought that the initial phases of imposex corresponding to VDS stages 1 and 2 may be reversible; however, advanced phases of imposex and sterilization with gross morphological changes corresponding to VDS stage 5 and 6 would be irreversible (Fent 1996).

Although many morphological aspects of pseudohermaphodi(ti)sm caused by TBT have been investigated, the biochemical mechanism has been indistinct. It is known that a neurotropic hormone called the penis morphogenic factor (PMF) develops male normal differentiation in mollusks (Féral and LeGall, 1983). Co-localization of TBT with PMF in ganglia suggested that PMF release through TBT's neurotropic action induced masculinization in females (Bryan et al. 1989). Other studies indicated increased testosterone levels detected in female dogwhelk exposed to TBT, and that testosterone injection without TBT induced penis development in females (Spooner et al. 1991, Stroben et al. 1991). The later studies suggested that TBT disturbed the P-450–dependent aromatization of androgens to estrogen, and a nonsteroidal specific aromatase inhibitor–induced imposex similar to TBT (Bettin et al. 1996). However, the PMF has not been well characterized, and the role of vertebrate sex steroids is not known in gastropods to date. A recent study proposed that the combination of changes in the neuropeptide (APGWa-mide), which is considered to be a PMF in mud snails, and steroid hormones would lead to imposex induction at extremely low doses of TBT (Oberdörster and McClellan-Green 2002).

Effects on Fish

TBT or TPT exposure in early life stages induces altered embryonic development, and delayed or inhibited hatching in fish. Exposure of TBT or TBT to minnow eggs and larvae at concentrations of 0.2 to 18 µg/L in the water in which the fish lived induced dose-dependent morphological effects on larvae. Marked body axis deformations were observed at more than about 4 µg/L exposure, and incomplete hatching occurred at similar concentrations in 10 to 30% of larvae. At 15.9 µg/L of TPT exposure, hatching was delayed and the hatching rate was reduced significantly (Fent and Meier 1992, 1994). Developmental defects, such as skeletal abnormality and retarded yolk sac resorption, occurred in zebrafish larvae at more than 25 µg/L of triphenyltin acetate (TPTA) exposure, and hatching delay was found at more than 0.5 µg/L (Strmac and Braunbeck 1999). These developmental effects in fish were caused not only by organotin compounds, but also by a variety of contaminants (i.e., heavy metals, chlorinated hydrocarbons, altered pH), suggesting that such alteration would be classified as a nonspecific reaction to organic toxicants (Fent 1996, Strmac and Braunbeck 1999).

Some reproductive effects (i.e., reduced fecundity and sperm counts) in fish were reported. Reproductive success of three-spine stickleback with TBT exposure were examined over a 7-month period; no effects were detected in relation to fecundity, number of hatched fry, or frequency of malformed fry. However, no changes were found in the gonad somatic index (GSI; ovary weight ratio to total body weight); by the 7-month TBT treatment (2 µg/L) despite increasing GSI in controls, which suggested a lack of maturation of egg tissue and consequently a potential reduced fecundity (Holm et al. 1991). In sheepshead minnows, reduction in both total and percent viable eggs was found at more than 1.3 µg/L of TBT exposure, although the reductions were not statistically significant (Manning et al. 1999). TBT exposure to Japanese medaka at 1 mg/kg body weight caused a reduction of the spawning frequency (Nirmala et al. 1999). Additionally, environmentally relevant concentrations of

TBT induced significantly decreased sperm counts in guppies (11.2 to 22.3 ng/L for 21 days), and decreased sperm motility at concentrations less than 1 µg/L (Haubruge et al. 2000, Kime et al. 2001).

Effects on Other Organisms

Despite a great number of studies on imposex in snails and a comparable number of toxicity reports on fish, there is little information on development and reproductive effects on other species by organotin compounds. It was reported that imposex has not only been found in gastropods, but also been induced in Japanese freshwater crabs by TBT (Takahashi et al. 2000). In crabs, imposex has also occurred in males, which is characterized by dual-gender imposex (either a female genital opening or a single ovary occurred in males). Malformations during limb regeneration occurred in fiddler crabs (Weis and Kim 1988) and in axolotl, induced by TBT (Scadding 1990).

Summary of Effects on Aquatic Organisms

TBT or TPT causes the imposition of male sex organs (imposex) on female mud snails above the concentration of about 1 ng/L (Sn) in seawater, but DBT or MPT does not induce imposex. The intensity is characterized by a classification system based on the VDS, and advanced phases of imposex and sterilization with gross morphological changes are irreversible. The biochemical mechanism studies suggested that the induction of either neu-rotropic hormone or androgen titer would lead to imposex at an extremely low dose of TBT. Also, TBT or TPT exposure in the early life stages of fish causes altered embryonic development, impaired morphological develop-ment, and delayed or inhibited hatching, and reduces fecundity and sperm counts. Such reproductive and developmental defects were also found in other species. The impaired reproduction and subsequent population decline in a variety of aquatic organisms by organotins are an important issue in aquatic ecosystems.

Effects on Experimental Animals

Reproductive Toxicity of Phenyltin Compounds

Reproductive Toxicity of Triphenyltins

TPTs have been reported to be insect chemosterilants (Kenaga 1965). Repro-ductive studies on TPTs are presented in Table 3.1. Several reports on male reproductive toxicity have been published. Male Sharman rats were given a diet containing triphenyltin hydroxide (TPTH) at 50, 100, or 200 ppm and then mated with untreated females repeatedly five times (Gaines and Kim-brough 1968). Reduced fertility, such as decreases in the total number of matings, total number of litters born alive, and ratio of number of litters to number of matings, accompanied by a marked reduction in food consumption

Table 3.1 Reproductive Toxicity of Phenyltin Compounds

Compounds	Animals	Dose	Days of Administration	Route	Reproductive and Developmental Effects	Author(s)
TPTH	Sharman rat	100–200 ppm	64–238 days	Diet	Decreased no. of matings Decreased no. of litters born alive Decreased ratio of no. of litters to no. of matings	Gains and Kimbrough (1968)
TPTA or TPTCl	Holtzman rat	20 mg/kg	19 days	Diet	Decreased tesicular size Change in testicular morphology	Pate and Hays (1968)
TPTA or TPTCl	Holtzman rat	20 mg/kg	20 days	Diet	Impairment of spermatogenic process	Snow and Hays (1983)
TPTA	ICR/Ha Swiss mouse	2.4–12 mg/kg 6 mg/kg	1 day 5 days	ip Gavage	No dominant lethal effect No dominant lethal effect	Epstein et al. (1972)
TPTH		1.3–8.5 mg/kg 11 mg/kg	1 day 5 days	ip Gavage	No dominant lethal effect No dominant lethal effect	
TPTA or TPTCl	Holtzman rat	20 mg/kg	4–24 days	Diet	Decreased no. of mature follicles Increased incidence of atresia in early follicle growth Decreased no.of corpora lutea	Newton and Hays (1968)
TPTCl	Wistar rat	4.7–6.3 mg/kg 12.5–25 mg/kg	Days 0–3 of pregnancy Days 4–6 of pregnancy	Gavage Gavage	Decreased pegnancy rate, decreased fetal wt. Decreased pregnancy rate	Ema et al. (1997a)
TPTCl	Wistar rat	4.7–6.3 mg/kg	Days 0–3 of pseudopregnancy	Gavage	Suppression of uterine decidualization	Ema et al. (1999a)
DPTCl	Wistar rat	16.5–24.8 mg/kg	Days 0–3 of pregnancy	Gavage	Decreased pregnancy rate, preimplatation loss, decreased fetal wt.	Ema et al. (1999b)
DPTCl	Wistar rat	33.3 mg/kg	Days 4–7 of pregnancy	Gavage	Effects as above, postimplantation loss	
DPTCl	Wistar rat	4.1–24.8 mg/kg	Days 0–3 of	Gavage	Suppression of uterine decidualization	Ema and Miyawaki (2002)

and weight gain, were observed at 100 or 200 ppm for 64 days. At these doses, food consumption later improved, and with it, fertility. Dietary exposure to triphenyltin acetate (TPTA) or triphenyltin chloride (TPTCl) at 20 mg/kg for 19 days produced marked effects on body weight, testicle size, and testicular structure in male Holtzman rats (Pate and Hays 1968). Microscopic examinations revealed degenerative changes, such as a decrease in the number of layers per tubule, a depletion of the more advanced cell forms from the tubules, and a closing of the tubule lumina. Effects were more pronounced in rats treated with TPTA. TPTA or TPTCl at 20 mg/kg in feed for 20 days was reported to cause an impairment of the spermatogenic process in male Holtzman rats; complete recovery of the spermatogenesis was observed after feeding a normal diet for 70 days (Snow and Hays 1983). No mutagenicity was detected in dominant lethal assay in which male ICR/ Ha Swiss mice were given a single intraperitoneal injection of TPTA at 2.4 or 12 mg/kg or TPTH at 1.3 or 8.5 mg/kg, or given TPTA at 6mg/kg or TPTH at 11 mg/kg by gavage on 5 successive days and then mated with untreated females, and pregnancy outcome was determined on day 13 of pregnancy (Epstein et al. 1972).

Adverse effects on female reproductive toxicity were also reported. Dietary TPTA and TPTCl at 20 mg/kg for 4 days produced significant changes in the ovarian tissue, including a decreased number of mature follicles, an increased incidence of atresia in early follicle growth, and a pronounced decrease in the number of corpora lutea in female Holtzman rats (Newton and Hays 1968). These effects were regarded as a decrease in ovulation, and thus decreased fertility. The adverse effects of TPTCl on the initiation and maintenance of pregnancy were determined after administration to the mother during early pregnancy (Ema et al. 1997a). Following successful mating, female Wistar rats were given TPTCl by gavage on days 0 to 3 of pregnancy at 3.1, 4.7, or 6.3 mg/kg or on days 4 to 6 of pregnancy at 6.3, 12.5, or 25.0 mg/kg, and pregnancy outcome was determined on day 20 of pregnancy. TPTCl totally prevented implantation in a dose-dependent manner. The pregnancy rate was decreased after administration of TPTCl on days 0 to 3 at 4.7 and 6.3 mg/kg and on days 4 to 6 at 12.5 and 25.0 mg/ kg. Preimplantation loss was increased after administration of TPTCl on days 0 to 3 at 4.7 mg/kg and higher. In females having implantations, the numbers of implantations and live fetuses, and the incidences of pre- and postimplantation embryonic loss in the TPTCl-treated groups were comparable to the controls. These results indicate that TPTCl during early pregnancy causes failure in implantation and has greater antiimplantation effects when administered during the preimplantation period than the periimplantation period.

The function of the uterine endometrium is one of the principle factors in embryonic survival. Uterine decidualization is required for normal implantation, placentation, and therefore normal gestation in rats. The uterine growth induced by endometrial trauma in pseudopregnant animals mimics the decidual response of the pregnant uterus that occurs after embryo implantation (Cummings 1990, Kamrin et al. 1994). The decidual cell response (DCR) is a model for maternal physiological events that are associated with implantation (Cummings 1990). This technique can distinguish between the adverse effects of chemical compounds in the maternal and fetal compartments, and has been used to evaluate the reproductive toxicity of chemical compounds (Spencer and Sing 1982, Bui et al. 1986, Cummings 1990, Kamrin et al. 1994, Ema et al. 1998). The effects of TPTCl on the reproductive capability of the uterus, as a cause of implantation failure, were evaluated using pseudopregnant rats (Ema et al. 1999a). Female Wistar rats were given TPTCl by gastric intubation at 3.1, 4.7, or 6.3 mg/kg on days 0 to 3 of pseudopregnancy. Between 11:00 and 13:00 on day 4 of pseudopregnancy, induction of DCR was performed via midventral laparotomy under ether anesthesia, and experimental decidualization was initiated by scratching the antimesometrial surface of the endometrium with a bent needle. The uterine weight on day 9 of pseudopregnancy served as an index of the uterine decidualization (De Feo 1963). A decrease in the uterine weight, which indicates suppression of the uterine decidualization, was detected at 4.7 and 6.3 mg/kg. TPTCl at 4.7 and 6.3 mg/kg also produced a decrease in the serum progesterone levels in female rats on day 4 and on day 9 of pseudopregnancy. These doses caused an increase in implantation failure (preimplantation embryonic loss) in female rats given TPTCl on days 0 to 3 of pregnancy (Ema et al. 1997a). These results suggest that TPTCl causes the suppression of uterine decidualization correlated with the reduction in serum progesterone levels, and these participate in the induction of implantation failure due to TPTCl. Protective effects of progesterone against suppression of uterine decidualization and implantation failure induced by TPTCl were examined (Ema and Miyawaki 2001). The hormonal regimen, consisting of progesterone and estorone supported decidual development in ovariectomized rats given TPTCl. The pregnancy rate and number of implantations in groups given TPTCl at 4.7 or 6.3 mg/kg in combination with progesterone were higher than those in the groups given TPTCl alone. These results indicate that the TPTCl-induced suppression of uterine decidualization is mediated, at least partially, by ovarian hormones, and that progesterone protects against TPTCl-induced implantation failure.

Reproductive Toxicity of Diphenyltin Compounds

Oral TPT is metabolized to DPT, MPT, and further to inorganic tin in rats (Kimmel et al. 1977, Ohhira and Matsui 1993 a, b). Reproductive toxicity studies on DPTs are also published (Table 3.1). The adverse effects of diphenyltin dichloride (DPTCl) on the initiation and maintenance of pregnancy, and the role of DPT in the implantation failure of TPT were evaluated. Following successful mating, DPTCl was given to Wistar rats by gavage on days 0 to 3 of pregnancy at 4.1, 8.3, 16.5, or 24.8 mg/kg or on days 4 to 7 of pregnancy at 8.3, 16.5, 24.8, or 33.0 mg/kg (Ema et al. 1999b). The pregnancy rate was decreased after administration of DPTCl on days 0 to 3 at 24.8 mg/kg and on days 4 to 7 at 33.0 mg/kg. The incidence of preimplantation loss was increased at 16.5 mg (equivalent to 48 µmol)/kg on days 0 to 3. In females having implantations, the incidences of pre- and postimplantation embryonic loss in the groups given DPTCl on days 0-3 were comparable to the controls. The incidence of postimplantation embryonic loss was increased after administration of DPTCl on days 4 to 7 at 33.0 mg/kg. These results indicate that DPTCl during early pregnancy causes implantation failure, and that DPTCl has greater effects on reproduction when administered during the preimplantation period rather than the periimplantation period. Following administration on days 0 to 3 of pregnancy, the increased incidence of preimplantation embryonic loss was induced by TPTCl, a parent compound of DPTCl, at 4.7 mg (equivalent to12 μmol)/kg and higher (Ema et al. 1997a), or DPTCl at 16.5 mg (equivalent to 48 μmol)/kg. If, on a mole-equivalent basis, a metabolite is as, or more, effective than the parent compound, this is consistent with the view that the metabolite is the proximate toxicant or at least an intermediate to the proximate toxicant. Thus, it seems unlikely that only DPTCl and/or its further metabolites can be considered the agents responsible for the antiimplantation effects of TPTCl. As for the metabolism of phenyltin, however, Ohhira and Matsui (1993b) showed that TPT compound was formed in the liver of the DPTCl-treated rat by metabolism of DPTCl, and suggested that part of the administered DPT compound has some harmful effect as the TPT compound in rats, and this must be taken into consideration in toxicological research on DPT. Further studies are needed to clarify the difference in the reproductive toxicity induced by TPT and DPT, and to identify the proximate or ultimate toxicant of phenyltins. The effects of DPTCl on the reproductive capability of the uterus were evaluated in pseudopregnant rats according to the procedure described above. Female Wistar rats were given DPTCl by gastric intubation on days 0 to 3 of pseudopregnancy at 4.1, 8.3, 16.5, or 24.8 mg/kg (Ema and Miyawaki 2002). Suppression of uterine decidualization was observed at 16.5 mg/kg and higher. A decrease in the serum progesterone levels in pseudopregnant rats was also found on day 4 and on day 9 of pseudopregnancy at 16.5 mg/kg and higher. These doses induced an increase in preimplantation embryonic loss in female rats given DPTCl on days 0 to 3 of pregnancy (Ema et al. 1999b). No changes in serum estradiol levels in pseudopregnant rats were noted. These results suggest that DPTCl causes the suppression of uterine decidualization correlated with the reduction in serum progesterone levels. These are responsible for the DPTCl-induced implantation failures. The hormonal regimen consisting of progesterone and estrone supported decidual development in ovariectomized rats given DPTCl (Ema and Miyawaki 2002). The pregnancy rate and number of implantations in groups given DPTCl at 16.5 or 24.3 mg/kg in combination with progesterone were higher than those in the groups given DPTCl alone. These results show that the DPTCl-induced suppression of uterine decidualization is mediated, at least partially, by ovarian hormones, and that progesterone protects against the DPTCl-induced implantation failure.

Summary of Reproductive Toxicity of Phenyltin Compounds

TPTs caused a decrease in male fertility due to degenerative changes in testicular tissue, which were associated with a marked decrease in food consumption. Complete recovery of fertility and impairment of the spermatogenesis was noted following withdrawal of treatment. Female reproductive failure induced by TPTs is more prominent. The harmful effects of TPTs on the ovaries were present after 5 days of treatment, before any significant effects on body weight gain. TPTCl during early pregnancy caused implantation failure at relatively low doses, and TPTCl had greater antiimplantation effects when administered during the preimplantation period. The implantation failure due to TPTCl might be mediated by suppression of uterine decidualization and correlated with the reduction in serum progesterone levels. Implantation failure and suppression of uterine decidualization accompanied with decreased levels of serum progesterone were also observed in rats given DPT, a major metabolite of TPT.

Developmental Toxicity of Phenyltin Compounds

Table 3.2 presents the developmental toxicity studies on phenyltin compounds given to female animals during pregnancy. Several reports on the adverse effects of phenyltins on development of offspring following maternal exposure have been published. Female SD rats were given TPTA by gavage at 5, 10, or 15 mg/kg on days 6 to 15 of pregnancy (Giavini et al. 1980). TPTA caused a decrease in maternal body weight gain at 10 mg/kg and higher, an increase in postimplantation loss at 15 mg/kg, and a reduction of fetal ossification at 5 mg/kg and higher. Teratogenic effects of TPTA were not found even at doses resulting in clear maternal toxicity. Depression of maternal body weight gain and food intake at 9.0 mg/kg and higher, and increase in postimplantation embryonic loss and decrease in fetal ossification at 9.0 mg/kg and higher, but not teratogenic effects, were observed in Wistar rats after administration of TPTA at 1.5, 3.0, 6.0, 9.0, or 12.0 mg/kg by gavage on days 7 to 17 of pregnancy (Noda et al. 1991a). Behavioral effects of prenatal exposure to TPTA were reported. A transient increase in spontaneous locomotor activity and increased mortality during the lactation period were found in pups of CFY rats given TPTA by gavage at 6 mg/kg on days 6 to 14 of pregnancy (Lehotzky et al. 1982). In this study, maternal rats were free of any overt signs of toxicity. Disruptions of learning acquisition, as evidenced by low avoidance rate in the Sidman avoidance test, and prolonged swimming time to the goal, and an increased number of errors in a reversed test in the water E-maze, were observed in postnatal offspring of Tokai High Avoiders (THA) rats received TPTA by gavage on days 6 to 20 of pregnancy at 4 or 8 mg/kg (Miyake et al. 1991). Maternal deaths and decreased weight gain were found at 8 mg/kg, no maternal toxicity was observed at 4 mg/kg, and no malformed offspring appeared in any group.

Table 3.2 Developmental Toxicity of Phenyltin Compounds

Compounds	Animals	Dose	Days of Administration	Route	Reproductive and Developmental Effects	Author(s)
TPTA	Wistar rat	5–15 mg/kg	Days 6–15 of pregnancy	Gavage	Postimplantation loss, delayed ossification	Giavini et al. (1980)
TPTA	Wistar rat	9–12 mg/kg	Days 7–17 of pregnancy	Gavage	Postimplantation loss, delayed ossification	Noda et al. (1991a)
TPTA	CFY rat	6 mg/kg	Days 6–14 of pregnancy	Gavage	Postnatal death, transient increase in spontaneous locomotor activity	Lehotzky et al. (1982)
TPTA	THA rat	4–8 mg/kg	Days 6–20 of pregnancy	Gavage	Disruption of learning acquisition	Miyake et al. (1991)
TPTH	SD rat	20 mg/kg	Days 1–7 of pregnancy	Gavage	Decreased pregnancy rate	Winek et al. (1978)
		15 mg/kg	Days 8–14 of pregnancy	Gavage	Posimplantation loss, decreased fetal wt.	
TPTH	SD rat	15 mg/kg	Days 14–12 of pregnancy	Gavage	Effects as above	Chernoff et al. (1990)
TPTCl	Wistar rat	13 mg/kg	Days 6–15 of pregnancy	Gavage	Postimplantation loss	Ema et al. (1999c)
		6.3–12.5 mg/kg	Days 7–9 of pregnancy	Gavage	Postimplantation loss	
		9.4–12.5 mg/kg	Days 10–12 or 13–15 of pregnancy	Gavage	Postimplantation loss, decreased fetal wt.	

Winek et al. (1978) noted that (1) SD rats given Vancide KS (TPTH) by gavage at 20 mg/kg on days 1 to 7 of pregnancy did not produce pups nor did they exhibit any resorption sites, (2) that only two of the six rats given TPTH at 15 mg/kg on days 14 to 20 of pregnancy produced viable pups, and (3) that four of the six rats given TPTH at 15 mg/kg on days 14 to 20 of pregnancy produced viable pups. Their study was conducted on only a small number of animals and the design of the study was not described in detail. Chernoff et al. (1990) observed a significant decrease in maternal body weight gain and an increase in postimplantation embryonic loss, but not fetal malformations, after administration of TPTH by gavage on days 6 to 15 of pregnancy at 13 mg/kg in SD rats. They stated that there was a correlation between maternal toxicity and fetal weight and/or lethality.

Following administration of TPTCl by gavage to pregnant Wistar rats, the maternal body weight gain and food consumption were decreased at 3.1 mg/kg and higher on days 7 to 9 of pregnancy, and at 6.3 mg/kg and higher on days 10 to 12 or on days 13 to 15 of pregnancy (Ema et al. 1999c). An increase in the incidence of postimplantation embryonic loss was found in pregnant rats given TPTCl at 6.3 mg/kg and higher on days 7 to 9, and at 9.4 mg/kg and higher on days 10 to 12 and on days 13 to 15. A decreased fetal weight was observed at 12.5 mg/kg on days 10 to 12 and at 9.4 mg/kg and higher on days 13 to 15. No increase in the incidence of fetuses with malformations was detected after administration of TPTCl regardless of the days of administration. These results indicate that TPTCl is developmentally toxic and that TPTCl has greater embryolethal effects when administered during earlier than later stages of organogenesis.

Summary of Developmental Toxicity of Phenyltin Compounds

Maternal exposure to TPTs caused embryonic/fetal death and suppression of fetal growth at maternal toxic doses. TPTs may cause reduction of fetal ossification at doses that are nontoxic to the mother. TPTs did not induce an increased number of fetal malformations even at doses producing overt maternal toxicity. Behavioral changes were reported in postnatal offspring of maternal rats that received TPTs during pregnancy at doses that did not cause overt maternal toxicity.

Reproductive Toxicity of Butyltin Compounds

Table 3.3 shows reproductive toxicity studies on butyltins. A decrease in the sperm head count and vacuolization of Sertoli cells were found in ICR mice gavaged with TBTO at 2 and 10 mg/kg twice a week for 4 weeks (Kumasaka et al. 2002). The male reproductive toxicity of tributyltin chloride (TBTCl) was reported in a two-generation reproductive toxicity study using Wistar rats (Omura et al. 2001). F0 females were fed a diet containing TBTCl at 5, 25, or 125 ppm (estimated to be 0.4, 2.0, or 10.0 mg/kg) from day 0 of pregnancy to the day of weaning of F1 rats. Feeding of TBTCl was continued

Table 3.3 Reproductive Toxicity of Butyltin Compounds

Compounds	Animals	Dose	Days of Administration	Route	Reproductive and Developmental Effects	Author(s)
TBTO	ICR mouse	2–10 mg/kg	4 weeks (twice a week)	Gavage	Decreased sperm head count, vacuolization of Sertoli cells	Kumasaka et al. (2002)
TBTCl	Wistar rat	25–125 ppm	2 generations	Diet	Decreased wt of testis and epididymis, decreased spermatid count, decreased levels of serum estradiol, decreased wt. gain of male offspring	Omura et al. (2001)
TBTCl	Wistar rat	5–125 ppm	2 generations	Diet	Decreased birth index, decreased no. and wt. of pups, delayed vaginal opening, increased female AGD, decreased wt. gain of female offspring	Ogata et al. (2001)
TBTCl	Wistar rat	12.2–16.3 mg/kg	Days 0–7 of pregnancy	Gavage	Decreased pregnancy rate, decreased fetal wt.	Harazono et al. (1996)
TBTCl	Wistar rat	16.3–32.5 mg/kg	Days 0–3 of pregnancy	Gavage	Decreased pregnancy rate, decreased fetal wt.	Harazono et al. (1998b)
TBTCl	Wistar rat	16.3–65.1 mg/kg	Days 4–7 of pregnancy	Gavage	Effects as above, postimplantation loss	
TBTCl	Wistar rat	16.3–32.5 mg/kg	Days 0–3 of pseudopregnancy	Gavage	Suppression of uterine decidualization, decreased levels of serum progesterone, increased levels of serum estradiol	Harazono and Ema (2000)
	Wistar rat	16.3–65.1 mg/kg	Days 4–7 of pseudopregnancy	Gavage	Suppression of uterine decidualization, decreased levels of serum progesterone	
DBTCl	Wistar rat	7.6–15.2 mg/kg	Days 0–3 or 4–7 of pregnancy	Gavage	Decreased pregnancy rate, pre- and postimplantation loss, decreased fetal wt.	Ema and Harazono (2000)
DBTCl	Wistar rat	7.6–15.2 mg/kg	Days 0–3 or 4–7 of pseudopregnancy	Gavage	Suppression of uterine decidualization, decreased levels of serum progesterone	Harazono and Ema (2003)
MBTCl	Wistar rat	903 mg/kg	Days 0–3 or 4–7 of pregnancy	Gavage	Decreased fetal wt.	Ema and Harazono (2001)

throughout the premating, mating, gestation, and lactation periods, for two generations. TBTCl affected the male reproductive system. The effects of TBTCl in the F2 generation were greater than those in the F1 generation. Body weight gain was consistently suppressed at 125 ppm in F1 and F2 males. The weights of the testis and epididymis were decreased and homogenization-resistant spermatid and sperm counts were reduced mainly at 125 ppm. Ventral prostate weight and spermatid count were decreased at 125 ppm in F1 males and at 25 and 125 ppm in F2 males. The serum 17-estradiol levels were decreased at 125 ppm in F1 and F2 males, but serum levels of luteinizing hormone and testosterone were not decreased. Omura et al. (2001) note that these changes corresponded with those caused by aromatase inhibitor and suggest that TBTCl might cause a weak aromatase inhibition in male rats.

Regarding female reproductive toxicity, the results with female rats in the above-mentioned two-generation reproduction study were reported by Ogata et al. (2001). Decreases in body weight gain during pregnancy, total number and average body weight of pups, and live birth index were observed at 125 ppm in F0 and F1 dams. Body weight gain was consistently suppressed at 125 ppm in F1 and F2 females. Delayed vaginal opening and impaired estrous cyclicity were found at 125 ppm in F1 and F2 females. The normalized anogenital distance (AGD) was increased at 5 ppm and higher in F1 females on postnatal day 1, and at 125 ppm in F1 and F2 females on postnatal days (PNDs) 1 and 4. These results show that a whole-life exposure to TBTCl affects the sexual development and reproductive function of female rats. They noted that TBTCl-induced increase in female AGD seems to suggest that it may exert a masculinizing (androgenic) effect on female pups.

Female Wistar rats were administered TBTCl by gavage at 8.1, 12.2, or 16.3 mg/kg on days 0 to 7 of pregnancy, and the adverse effects of TBTCl on implantation and maintenance of pregnancy were determined (Hrazono et al. 1996). Decreases in maternal body weight gain at 12.2 mg/kg and higher, and food consumption at 8.1 mg/kg and higher, were found. Implantation failure was found at doses that also produced maternal toxicity. The pregnancy rate was significantly decreased at 12.2 mg/kg and higher. In females having implantations, the numbers of corpora lutea, implantations, and postimplantation loss, were comparable across all groups. To examine whether pregnancy failure was the result of the effects of TBTCl or maternal malnutrition from reduced food consumption, a pair-feeding study was performed. The results show that the pregnancy failure observed in the TBTCl-treated group is due to the effects of TBTCl, not to the maternal malnutrition from reduced food consumption (Harazono et al. 1998a). The adverse effects of TBTCl on implantation and maintenance of pregnancy after administration during the pre- or periimplantation period were evaluated. Female Wistar rats were given TBTCl by gastric intubation on days 0 to 3 of pregnancy at 4.1, 8.1, 16.3, or 32.5 mg/kg, or on days 4 to 7 of pregnancy at 8.1, 16.3, 32.5, or 65.1 mg/kg, and pregnancy outcome was determined on day 20 of pregnancy (Harazono et al. 1998b). TBTCl on days

0 to 3 at 16.3 mg/kg and higher and on days 4 to 7 at 65.1 mg/kg caused a decrease in pregnancy rate and an increase in preimplantation embryonic loss. TBTCl on days 4 to 7 of pregnancy caused a significant increase in the incidence of postimplantation loss at 16.3 mg/kg and higher. The results show that the manifestation of adverse effects of TBTCl varies with gestational stage at the time of maternal exposure, and that TBTCl during the preimplantation period causes implantation failure, while TBTCl during the periimplantation period adversely affects the viability of implanted embryos.

Female Wistar rats were given TBTCl by gavage on days 0 to 3 or on days 4 to 7 of pseudopregnancy, and the effects of TBTCl on the uterus, as a cause of implantation failure, were evaluated according to the same procedures described above. After administration of TBTCl on days 0 to 3 of pseudopregnancy, a decrease in the uterine weight was detected at 16.3 mg/kg and higher (Harazono and Ema 2000). Decreased levels of serum progesterone occurred on day 9 at 16.3 mg/kg and higher and on day 4 at 8.1 mg/kg and higher, and increased levels of serum estradiol at 32.5 mg/kg were observed after administration on days 0 to 3. Following administration of TBTCl on days 4 to 7 of pseudopregnancy, uterine weight and serum progesterone levels on day 9 decreased at 16.3 mg/kg and higher. The doses that induced decreases in uterine weight and serum progesterone levels in pseudopregnant rats are consistent with those that induced pre- and postimplantation loss in pregnant rats. These results indicate that TBTCl suppresses uterine decidualization correlated with a reduction in serum progesterone levels, and suggest that the decline in uterine decidualization and serum progesterone levels participate in the induction of implantation failure induced by TBTCl.

TBT compound is reported to be metabolized to di- and MBT derivatives, and DBT was metabolized to MBT in rats (Fish et al. 1976, Kimmel et al. 1977, Ishizaka et al. 1989, Iwai et al. 1981). The adverse effects of dibutyltin dichloride (DBTCl) on the implantation and maintenance of pregnancy, and the role of DBT in the reproductive toxicity of TBT were evaluated after maternal exposure during the pre- or periimplantation period (Ema and Harazono 2000). Female Wistar rats were given DBTCl by gastric intubation at 3.8, 7.6, or 15.2 mg/kg on days 0 to 3 or on days 4 to 7 of pregnancy. The pair-feeding study was also performed. After administration of DBTCl on days 0 to 3, the pregnancy rate in the 7.6 mg/kg group was lower than in the control group, and that in the 15.2 mg/kg group was lower than in the control and pair-fed groups. The incidence of postimplantation embryonic loss in the groups given DBTCl on days 4 to 7 at 7.6 and 15.2 mg/kg was higher than in the control and pair-fed groups. Early embryonic loss was considered to be due to the effects of DBTCl, not to maternal malnutrition from reduced feed consumption, and the lowest dose of DBTCl inducing early embryonic loss was conservatively estimated at 7.6 mg (25 µmol)/kg. An increase in the incidence of implantation failure was observed after administration of TBTCl, the parent compound of DBTCl, at 16.3 mg (50 µmol)/kg and higher on days 0 to 3 and on days 4 to 7 of pregnancy, respectively. The doses of DBTCl that caused early embryonic loss were lower than those of TBTCl (Harazono et al. 1998b). Thus, it is likely that DBTCl and/or its metabolites can be considered the agents responsible for early embryonic loss induced by TBTCl. Suppression of uterine decidualization accompanied by reduced levels of serum progesterone was found in pseudopregnant rats given DBTCl at doses that caused implantation failure (Harazono and Ema 2003), and administration of progesterone protected, at least in part, against the DBTCl-induced implantation failure (Ema et al. 2003). These results suggest that the decline in progesterone levels is a primary mechanism for the implantation failure due to DBTCl. Administration of butyltin trichloride (MBTCl) on days 0 to 3 or on days 4 to 7 of pregnancy did not cause pre- or postimplantation loss, even at 903 mg (equivalent to 3200 μmol)/kg in Wistar rats (Ema and Harazono 2001). It is unlikely that MBTCl and/or metabolites are actively involved in the early embryonic loss due to butyltins. The dose levels of DBTCl that suppressed the DCR were lower than the effective doses of TBTCl on a molar base. The similarity of effects and equivalent or greater effectiveness of DBTCl may suggest that DBTCl participates in the inhibition of DCR and in the decrease in serum progesterone levels associated with TBTCl. Although increased levels of serum estradiol on day 9 of pseudopregnancy was observed in rats given TBTCl on days 0 to 3 (Harazono and Ema 2000), administration of DBTCl did not affect serum estradiol levels. Thus, the mechanisms of TBTCl and DBTCl adversely affecting ovarian function might be different. Further studies are needed to determine the effects of TBTCl and DBTCl on the maternal endocrine system, including ovarian function.

Summary of Reproductive Toxicity of Butyltin Compounds

In a rat two-generation reproductive toxicity study, TBTCl affected the male and female reproductive system. TBTCl caused decreases in weight of the testis, epididymis, and ventral prostate, and spermatid and sperm counts in male offspring. The serum estradiol levels decreased in male offspring, but serum levels of luteinizing hormone and testosterone did not decrease. Total number and average body weight of pups, and live birth index decreased. Delayed vaginal opening and impaired estrous cyclicity were found in female offspring. The AGD increased even at 0.4 mg/kg in female offspring. TBTCl during early pregnancy caused implantation failure in rats. Implantation failure due to TBTCl may be mediated via the suppression of uterine decidualization and correlated with the reduction in serum progesterone levels. Implantation failure was also observed following administration of DBTCl, at lower doses than TBTCl, during early pregnancy. Suppression of uterine decidualization, accompanied by reduced levels of serum progesterone, was also observed in pseudopregnant rats given DBTCl at doses that induced implantation failure. Administration of progesterone protected, at least in part, against the DBTCl-induced implantation failure. Administration of MBTCl during early pregnancy did not cause pre- or postimplantation loss even at 903 mg/kg. These results suggest that DBT may be responsible for the TBT-induced implantation failure, and that the decrease in serum progesterone levels may be a primary factor in implantation failure due to butyltins.

Developmental Toxicity of Butyltin Compounds

In Vivo *Developmental Toxic Effects of Butyltin Compounds*
Studies on developmental toxicity of butyltins are shown in Table 3.4. Several studies concerning the developmental toxicity of TBTO have been conducted in mice and rats. Davis et al. (1987) reported that an increased incidence of resorptions, reduced fetal weight, and an increased incidence of cleft palate were accompanied by a marked decrease in maternal weight gain after administration of TBTO by gavage to NMRI mice on days 6 to 15 of pregnancy. TBTO at 11.7 mg/kg was the lowest dose resulting in reduced maternal weight with no indication of decreases in litter size and fetal weight. At 35 mg/kg, the incidence of resorptions was 59% and fetal weight was markedly lowered. Doses lower than 11.7 mg/kg did not cause clear-cut teratogenic effects, and the incidences of cleft palate were 7% at 11.7 mg/kg and 48% at 35 mg/kg. They concluded that cleft palate might be a nonspecific toxic effect and not a teratogenic effect of TBTO. Swiss albino mice received TBTO by gavage at on days 6 to 15 of pregnancy (Baroncelli et at. 1990, 1995). In the prenatal study, a decrease in maternal weight gain and fetal weight were found, along with high embryolethality, but no increased incidence of fetal malformations were found at 40 mg/kg (Baroncelli et al. 1990). In the postnatal study, reduced litter size and pup weight at 20 mg/kg and higher, increased percentage of dams that had not built a nest at 10 mg/kg and higher, and decreased maternal weight gain and increased number of early or late deliveries at 5 mg/kg were detected. No malformations in pups were observed (Broncelli et al. 1995). Nonspecific alterations of hematological parameters and thymus or spleen weights were noted in dams and offspring of Swiss mice after administration of TBTO by gavage at 5, 10, or 20 mg/kg on days 6 to 15 of pregnancy (Karrer et al. 1995). A high incidence of cleft palate (11.4 percent) was found at 27 mg/kg in Han:NMRI mice given TBTO by gavage on days 6 to 17 of pregnancy (Faqi et al. 1997). At this dose, two fetuses exhibited a bent radius, eight fetuses were observed with a short mandible, and five fetuses showed a fusion of the occipital bones with their basal parts. In this study, no signs of toxicity in maternal and fetal mice were detected up to the dose of 13.5 mg/kg. Long Evans rats were given TBTO by gavage at 2.5, 5, 10, 12, or 16 mg/kg on days 6 to 20 of pregnancy, allowed to give birth, and pups were examined (Crofton et al. 1989). Maternal body weight gain, and pup litter size, weight, and viability on PNDs 1 and 3 were decreased at 10 mg and higher. A 3% incidence of cleft palate was detected at 12 mg/kg. There were no pups born with malformations at 10 mg/kg and lower. Vaginal opening was delayed in females exposed to 10 mg/kg. Motor activity was decreased on PND 14 at all doses. Adult brain weight

Table 3.4 Developmental Toxicity of Butyltin Compounds

Compounds	Animals	Dose	Days of Administration	Route	Reproductive and Developmental Effects	Author(s)
TBTO	NMRI mouse	11.7–35 mg/kg	Days 6–15 of pregnancy	Gavage	Postimplantation loss, decreased fetal wt., cleft palate	Davis et al. (1987)
TBTO	Swiss mouse	40 mg/kg	Days 6–15 of pregnancy	Gavage	Postimplantation loss, decreased fetal wt.,	Baroncelli et al. (1990)
TBTO	Swiss mouse	10–30 mg/kg	Days 6–15 of pregnancy	Gavage	Decreased litter size, decreased pup wt., changed length of gestation, decreased percentage of dams exhibiting nest-building	Baroncelli et al. (1995)
TBTO	Swiss mouse	5–20 mg/kg	Days 6–15 of pregnancy	Gavage	Nonspecific effects on hematological parameters	Karrer et al (1995)
TBTO	Ha:NMRI mouse	27 mg/kg	Days 6–17 of pregnancy	Gavage	Decreased fetal wt., cleft palate, skeletal malformations	Faqi et al. (1997)
TBTO	Long Evans rat	2.5–16 mg/kg	Days 6–20 of pregnancy	Gavage	Decreased litter size and pup wt., cleft palate, decreased postnatal wt. gain, delayed vaginal opening, decreased brain wt., transient decrease in motor activity	Crofton et al. (1989)
TBTO	THA rat	5–10 mg/kg	Days 6–20 of pregnancy	Gavage	Postnatal death, disruption of learning acquisition	Miyake et al. (1990)
TBTA	Wistar rat	16 mg/kg	Days 7–17 of pregnancy	Gavage	Postimplantation loss, cleft palate, decreased fetal wt.	Noda et al. (1991b)
TBTCl	Wistar rat	5–25 mg/kg	Days 7–15 of pregnancy	Gavage	Postimplantation loss, delayed ossification	Itami et al. (1990)
TBTCl	Wistar rat	25–50 mg/kg	Days 7–9 of pregnancy	Gavage	Postimplantation loss, decreased fetal wt.	Ema et al. (1995a)

Table 3.4 Developmental Toxicity of Butyltin Compounds (continued)

Compounds	Animals	Dose	Days of Administration	Route	Reproductive and Developmental Effects	Author(s)
TBTCl	Wistar rat	50–100 mg/kg 25–100 mg/kg 100–200 mg/kg	Days 10–12 of pregnancy Days 13–15 of pregnancy One day during days 7–15 of pregnancy	Gavage Gavage Gavage	Effects as above, cleft palate Decreased fetal wt., cleft palate Postimplantation loss, decreased fetal wt., cleft palate after po on day 8, 11, 12, 13, or 14	Ema et al. (1997b)
TBTCl	SD rat	0.25–20 mg/kg	Days 0–19 of pregnancy	Gavage	Postimplantation loss, decreased fetal wt., increased male AGD, delayed ossification, decreased levels of serum thyroxine and triiodothyronine	Adeeko et al. (2003)
		2.5–10 mg/kg	Days 8–19 of pregnancy		Decreased levels of serum thyroxine	
TBTCl	SD rat	0.025–2.5 mg/kg	From day 8 of pregnancy until adulthood	Gavage	Decreased wt of liver, spleen and thymus, reduced serum levels of creatinine, triglyceride, amylase and thyroxine, change in growth profiles	Cooke et al. (2004)
TBTCl	SD rat	0.25–2.5 mg/kg	From day 8 of pregnancy until adulthood	Gavage	Thymus atrophy, increased no. of natural killer cells, increased levels of IgM and IgG, increased no. of immature T lymphocytes, decreased levels of IgG2a	Tryphonas et al. (2004)

Compound	Species	Dose	Timing	Route	Effects	Reference
TBTCl	SD rats	1–5 mg/kg	Days 6–20 of pregnancy	Gavage	Increased spontaneous activity, retarded aquisition of the radial arm maze task, potentiation of d-amphetamine-induced hyperactivity	Gårdlung et al. (1991)
DBTA	Wistar rats	15 mg/kg	Days 0–19 of pregnancy	Gavage	Postimplantation loss, decreased fetal wt, manibular dysplasia, ankyloglossia, schistoglossia, skeletal variation	Nada et al. (1988)
DBTA	Wistar rat	5–15 mg/kg	Days 7–17 of pregnancy	Gavage	Postimplantation loss, decreased fetal wt., cleft mandible, cleft lower lip, ankyloglossia, schistoglossia, tail anomaly, deformity of ribs and vertebrae, skeletal variations	Noda et al. (1992a)
MBTCl	Wistar rat	50–400 mg/kg	Days 7–17 of pregnancy	Gavage	No effects	Noda et al. (1992b)
DBTA	Wistar rat	15 mg/kg	Days 7–9 of pregnancy	Gavage	Effects as above	
DBTA	Wistar rat	22 mg/kg	Day 8 of pregnancy	Gavage	Malformations as above	Noda et al. (2001)
DBTCl	Wistar rat	10–22 mg/kg	Day 8 of pregnancy	Gavage	Malformations as above	Ema et al. (1991)
DBTCl	Wistar rat	5–10 mg/kg	Days 7–15 of pregnancy	Gavage	Postimplantation loss, decreased fetal wt, cleft jaw, cleft palate, ankyloglossia, omphalocere, tail anomaly, deformity of ribs and vertebrae	
DBTCl	Wistar rat	20 mg/kg	Days 7–9, 10–12, or 13–15 of pregnancy	Gavage	Decreased fetal wt., postimplantation loss, malformations as above after p.o. on days 7–9	Ema et al. (1992)

Table 3.4 Developmental Toxicity of Butyltin Compounds (continued)

Compounds	Animals	Dose	Days of Administration	Route	Reproductive and Developmental Effects	Author(s)
		20–40 mg/kg	Day 6, 7, 8, or 9 of pregnancy	Gavage	Decreased fetal wt., postimplantation loss after p.o. on day 6, 7, or 8, malformations as above after p.o. on day 7 or 8	Farr et al. (2001)
DBTCl	Wistar rat	1–10 mg/kg	Days 6–15 of pregnancy	Gavage	No effects	
DBTA	Wistar rat	28.1 mg/kg	Day 8 of pregnancy	Gavage	Cleft mandible, cleft lower lip, ankyloglossia, schistoglossia, exencephaly, deformity of ribs and vertebrae	Noda et al. (1993)
DBTCl	Wistar rat	24.3 mg/kg	Day 8 of pregnancy	Gavage	Decreased fetal wt., malformations as above	
DBTM	Wistar rat	27.8 mg/kg	Day 8 of pregnancy	Gavage	Malformations as above	
DBTO	Wistar rat	19.9 mg/kg	Day 8 of pregnancy	Gavage	Malformations as above	
DBTL	Wistar rat	50.0 mg/kg	Day 8 of pregnancy	gavage	Malformations as above	
3-OHDBTL	Wistar rat	100 mg/kg	Day 8 of pregnancy	Gavage	Decreased fetal wt., peaked mandible	
TeBT	Wistar rat	1832 mg/kg	Days 13–15 of pregnancy	Gavage	Cleft palate	Ema et al. (1996a)
TBTCl	Wistar rat	54–108 mg/kg	Days 13–15 of pregnancy	Gavage	Decreased fetal wt., cleft palate	
DBTCl	Wistar rat	50–100 mg/kg	Days 13–15 of pregnancy	Gavage	Decreased fetal wt.	
TBTCl	Wistar rat	40–80 mg/kg	Days 7–8 of pregnancy	Gavage	Postimplantation loss, decreased fetal wt.	Ema et al. (1995b)
DBTCl	Wistar rat	10–15 mg/kg	Days 7–8 of pregnancy	Gavage	Effects as above, malformations as above	
MBTCl	Wistar rat	1000–1500 mg/kg	Days 7–8 of pregnancy	Gavage	Decreased fetal wt.	

was reduced at 10 mg/kg. THA rats were given TBTO by gavage at 5 or 10 mg/kg on days 6 to 20 of pregnancy and allowed to deliver spontaneously, and pups were examined (Miyake et al. 1990). All pups died by PND 3 at 10 mg/kg. In pups at 5 mg/kg, prenatal TBTO disrupted learning acquisition in the Sidman avoidance test and a reversal test in the water E-maze.

Pregnant Wistar rats were given tributyltin acetate (TBTA) by gavage at 1, 2, 4, 8, or 16 mg/kg on days 7 to 17 of pregnancy (Noda et al. 1991b). An increase in incidences of intrauterine deaths, cleft palate, and low fetal weight were found at 16 mg/kg. This dose level also induced severe reductions in maternal weight gain and food consumption, and TBTA at 4 mg/ kg and higher lowered maternal thymus weight. Noda et al. (1991b) concluded that the observed teratogenic effects may not be a specific action of TBTA because their results were similar to those of Davis et al. (1987).

No live fetuses were obtained in Wistar rats treated with TBTCl at 25 mg/kg by gavage on days 7 to 15 of pregnancy (Itami et al. 1990). Maternal toxicity at 9 mg/kg and higher, and skeletal retardation in fetuses at 5 mg/ kg and higher were observed, but fetal malformations were not found. An increase in placental weight was found at 5 mg/kg and higher. To obtain more precise information on the effects of TBTCl on fetal development, Wistar rats were given TBTCl by gavage at relatively high doses during a shorter period, at 25 or 50 mg/kg on days 7 to 9, at 50 or 100 mg/kg on days 10 to 12, or at 25, 50, or 100 mg/kg on days 13 to 15 of pregnancy (Ema et al. 1995a). A decrease in maternal weight gain was observed in all groups regardless the days of administration. An increase in incidence of postimplantation embryonic loss was found in pregnant rats given TBTCl on days 7 to 9 at 25 mg/kg and higher, and on days 10 to 12 at 100 mg/kg, but not in pregnant rats given TBTCl on days 13 to 15 at up to 100 mg/kg. A lower fetal weight was observed in pregnant rats given TBTCl on days 10 to 12 at 50 and 100 mg/kg, and on days 13 to 15 at 100 mg/kg. An increased incidence of fetuses with malformations was detected after administration of TBTCl on days 10 to 12 at 100 mg/kg, and on days 13 to 15 at 25 mg/kg and higher. The most predominant malformation was cleft palate. These results indicate that the manifestation of abnormal development induced by TBTCl varies with developmental stage at the time of administration, and that TBTCl has teratogenic potential with developmental phase specificity. The most susceptible day to the teratogenicity of TBTCl was determined by a single administration on one of the days during organogenesis (Ema et al. 1997b). An increase in incidence of fetuses with external malformations was detected when TBTCl was given on day 8 at 100 and 200 mg/kg, or on day 11, 12, 13, or 14 at 200 mg/kg, and the most pronounced effect was seen after administration on day 13 of pregnancy. Cleft palate was mainly observed after administration of TBTCl. These findings indicate that TBTCl has a biphasic teratogenicity on day 8 and on days 11 to 14 of pregnancy. Pregnant SD rats were gavaged with TBTCl at 0.25, 2.5, 10, or 20 mg/kg on days 0 to 19 of pregnancy, or at 0.25, 2.5, or 10 mg/kg on days 8 to 19 of pregnancy, and pregnancy outcome was assessed (Adeeko et al. 2003). A reduced maternal weight gain, decrease in pregnancy rate, increase in postimplantation loss, and decrease in fetal weight were found after administration of TBTCl at 20 mg/kg on days 0 to 19. These findings support the previous results (Harazono et al. 1996, 1998a, b) in which TBTCl during early pregnancy at 12.2 mg/kg and higher caused increases in pre- and postimplantation loss. The incidence of fetal malformations was not increased in any TBTCl-treated groups. An increase in normalized AGD of male fetuses was detected at 0.25 mg/kg and higher on days 0 to 19, but not at any dose on days 8 to 19, even at the highest dose level. Hormonally active agents are known to affect mammalian internal and external genitalia when administered during sex differentiation (i.e., the perinatal period) (Schardein 2000). It is reported that days 16 to 17 of pregnancy were the most sensitive for finasteride-induced feminizing effects, including a decrease in AGD in male rat offspring (Clark et al. 1993), and that the period of days 15 to 17 of pregnancy was the most susceptible for dibutyl phthalate–induced decrease in the AGD of male rat offspring (Ema et al. 2000). There are discrepancies in the effects of TBTCl on the AGD between outcomes after exposure on days 0 to 19 and on days 8 to 19 of pregnancy, and between this study and previous reports in which whole-life exposure to TBTCl caused increase in female AGD in rat two-generation reproductive studies (Ogata et al. 2001). *In vitro* studies showed that TPT and TBT had an ability to activate androgen receptor mediated transcription in mammalian cells (Yamabe et al. 2000). TPTCl, TBTCl, and DBTCl caused aromatase inhibition in the human adenocortical carcinoma cell line (Sanderson et al. 2002). Although TeBT and MBTCl had no effect on either human 5-reductase type 1 or type 2, TBTCl and DBTCl influenced the human 5-reductase isozymes (Doering et al. 2002). DBTCl specifically inhibited brain 5-reductase type 1 with no effect on prostate 5-reductase type 2. TBTCl inhibited both isoenzymes. Doering et al. (2002) noted that the inhibition of the TBTCl inhibited both isoenzymes. Type 2 could potentially disturb normal male physiology. These *in vitro* findings may explain the *in vivo* reproductive and developmental outcomes induced by organotins. Adeeko et al. (2003) also noted that reduced fetal ossification of the sternebrae was found at 10 mg/kg and higher, for which fetal weights at 10 mg/kg were in normal range, and TBTCl at 10 mg/kg and higher during pregnancy decreased maternal circulating thyroid hormone levels and increased the weight of the placenta. They noted that the TBTCl-induced disturbances in maternal thyroid hormone homeostasis could contribute to the reduction in fetal skeletal ossification.

Pregnant SD rats were gavaged with TBTCl at 0.025, 0.25, or 2.5 mg/kg from day 8 of pregnancy until weaning, and offspring were gavaged with the same dose of TBTCl given to their mothers until adulthood (Cooke et al. 2004, Tryphonas et al. 2004). No effects of TBTCl on body weight, food consumption, or histopathological findings in the thyroid, liver, adrenal or colon were observed in maternal rats. No effects of TCBTCI on litter size, sex ratio, postnatal survival rate, or histopathological findings in the liver, adrenal glan or colon were also found in offspring. Decreased serum levels of creatinine, triglycerides, and magnesium in female offspring and of thyroxine in male offspring were found at 2.5 mg/kg. Decreased weight of the spleen in male offspring and the thymus in female offspring were observed at 0.25 mg/kg. Significant effects on growth profiles in male and female offspring, and decreased liver weights in female offspring were noted even at 0.025 mg/kg (Cooke et al. 2004). Immunotoxic effects of TBTCl were determined in these rat offspring (Tryphonas et al. 2004). Thymus atrophy, an increase in the number of natural killer cells and immunoglobulin M (IgM) levels, a decrease in the IgG2a levels at 2.5 mg/kg, and an increase in the mean percentage immature T lymphocytes and IgG levels at 0.25 mg/kg and higher were observed in offspring. Significant effects were found more frequently at 0.25 mg/kg and higher, and minor effects were observed at 0.025 mg/kg. Tryphonas et al. (2004) concluded that the low levels of TBTCl affected humoral and cell-mediated immunity, and the number and function of cells involved in the hostís immunosurveillance mechanisms against tumors and vital infections in rat offspring.

Postnatal behavioral changes in pups of SD rats that received TBTCl prenatally on days 6 to 20 of pregnancy, at doses not toxic to the mother, were also reported (Gårdlund et al. 1991). An increase in spontaneous activity, such as locomotion, rearing, and total activity, retarded acquisition in radial arm maze performance, and potentiation of d-amphetamine-induced hyperactivity were observed at 1 and 5 mg/kg.1

The adverse effects of DBT, a major metabolite of TBT, on embryonic/fetal development were assessed after maternal administration during organogenesis. Pregnant Wistar rats were given DBTA by gavage at 1.7, 5, or 15 mg/kg during the whole period, on days 0 to 19, of pregnancy (Noda et al. 1988). At 15 mg/kg, a decrease in body weight gain and thymus weight in dams, and a low body weight and increased number of fetal malformations occurred. Administration of DBTA by gavage during the organogenetic period, on days 7 to 17, of pregnancy at 10 mg/kg and higher also caused increased fetal malformations, such as cleft mandible, cleft lower lip, ankyloglossia, schistoglossia, exencephaly, anury, vestigial tail, and deformity of the ribs and vertebrae (Noda et al. 1992a). Decreases in thymus weight and fetal weight at 10 mg/kg and higher, and decreases in maternal weight gain at 15 mg/kg were observed following administration of DBTA on days 7 to 17 of pregnancy. The most susceptible gestational day to teratogenicity of DBTA in rats was day 8 of pregnancy (Noda et al. 1992b). Occurrences of similar types of fetal malformations after administration of DBTA on day 8 of pregnancy were also reported in other papers (Node et al. 1993, 1994, 2001). Teratogenic effects of DBTCl were also studied in Wistar rats. Female rats were given DBTCl by gavage at 2.5, 5.0, or 7.5 mg/kg on days 7 to 15 of pregnancy (Ema et al. 1991). The incidence of fetal malformations was increased and roughly proportional to the dose of DBTCl administered at 5.0 mg/kg and higher. Cleft jaw, ankyloglossia, omphalocele, anomaly of the tail, defect of the mandible, deformity of the vertebral column and ribs, and microphthalmia were frequently observed. In this study, decreases in maternal weight gain and food consumption was observed at 7.5 mg/kg and higher. These results indicate that DBTCl produce teratogenic effects in the absence of overt maternal toxicity. However, the thymus weight was not determined. The susceptible gestational days to teratogenicity of DBTCl was determined after administration of relatively high doses of TBTCl on days 7 to 9, on days 10 to 12, or on days 13 to 15 of pregnancy (Ema et al. 1992). An increase in fetal malformations and postimplantation loss was detected after administration of DBTCl at 20 mg/kg on days 7 to 9, but neither was detected on days 10 to 12 nor on days 13 to 15. The data of the study in which pregnant rats were given a single dose of DBTCl by gavage showed that developing offspring were not susceptible to teratogenicity of DBTCl on day 6, and that day 7 was the earliest susceptible period, day 8 was the most susceptible period, and day 9 was no longer a susceptible period with respect to the teratogenicity of DBTCl (Ema et al. 1992). Occurrences of similar types of fetal malformations after administration of DBTCl on day 8 or on days 7 to 8 of pregnancy were also reported in rats (Noda et al. 1993, Ema et al. 1995b). Farr et al. (2001) also reported the developmental toxicity of DBTCl in rats. Wistar rats were administered DBTCl by gavage at 1, 2.5, 5, or 10 mg/kg on days 6 to 15 of pregnancy. Decreases in maternal weight gain, food consumption, and thymus weight, but not developmental indicators, were observed at the highest dose tested, 10 mg/kg. At this dose, four fetuses out of 262 fetuses had malformations, including ankyloglossia, mandible defects, tail anomaly, and deformity of the vertebrae, which were similar types of malformations to those previously reported after administration of DBTA (Noda et al. 1988, 1992a, b, 1993, 1994, 2001) and DBTCl (Ema et al. 1991, 1992, 1995b, Noda et al. 1993). They concluded that a slightly increased, but not statistically significant, number of malformations was associated with the onset of maternal toxicity, and that no increase in developmental defects was induced at dose levels that did not result in maternal toxicity.

The teratogenic effects of five DBTs with different anions, such as DBTA, DBTCl, dibutyltin maleate (DBTM), dibutyltin oxide (DBTO), and dibutyltin dilaurate (DBTL), were determined in Wistar rats given by gavage at 80 µmol/kg on the most susceptible day for teratogenicity of DBTA and DBTCl (Noda et al. 1993). Although the incidences of fetuses with malformations were different among DBTs, the types of malformations induced by these DBTs are similar to those in the previous studies with DBTA. Noda et al. (1993) suggest the importance of the dibutyl group rather than the anionic group in the production of fetal malformations. They also noted that butyl(3-hydroxybutyl)tin dilaurate (3-OHDBL), one of the main metabolites of DBTCl (Ishizaka et al. 1989), was not responsible for the teratogenicity of DBTCl because of weak potential for production of fetal malformations.

TeBT is metabolized to tri-, di-, and monobutyltin derivatives (Kimmel et al. 1977). The TBT compound is metabolized to di- and monobutyltin derivatives, and DBT was metabolized to MBT in rats (Iwai et al. 1981). TeBT, TBTCl, DBTCl, and MBTCl were compared for their developmental toxicity to evaluate these butyltin compounds as potential toxicants in teratogenicity following administration of relatively high doses of butyltins to pregnant rats during the susceptible period to teratogenesis of TBTCl or during the susceptible period to teratogenesis of DBTCl. Pregnant rats were given TeBT, TBTCl, or DBTCl during the period of susceptibility to the teratogenesis of TBTCl, on days 13 to 15 of pregnancy (Ema et al. 1996a). TeBT caused an increased incidence of cleft palate at 1832 mg (5280µmol)/kg. TBTCl induced a markedly increased incidence of fetuses with cleft palate at 54 mg (165 µmol)/kg and higher, and decreased fetal weight at 108 mg (330 µmol)/kg. Following administration of DBTCl on days 13 to 15 of pregnancy, fetal weight was reduced at 54 mg (165 µmol)/kg and higher, but neither increase in postimplantation loss nor fetuses with malformations was found even at 100 mg (330 µmol)mg/kg. These results indicate that there are differences in the manifestation and degree of developmental toxicity among TeBT, TBT, and DBT. Pregnant rats received TBTCl, DBTCl, or MBTCl during the period of susceptibility to teratogenesis of DBTC, on days 7 to 8 of pregnancy (Ema et al. 1995b). TBTCl at 40 and 80 mg/kg caused an increase in postimplantation embryolethality, but no increase in fetal malformations. DBTCl caused a markedly high incidence of fetal malformations, lower fetal weight, and higher postimplantation embryonic loss at 10 mg/kg and higher. No increase in the incidences of postimplantation loss or malformed fetuses was observed after administration of MBTCl even at 1500 mg/kg. These results indicate that the developmental toxicity of DBTCl is different from that of TBTCl and MBTCl in the level of susceptibility and spectrum of toxicity. A lack of developmental toxicity of MBTCl was also reported by Noda et al. (1992a). MBTCl on days 7 to 17 of pregnancy did not affect maternal body weight and thymus weight, or fetal survival, growth, and morphological development, even at 400 mg/kg in Wistar rats. Their observations support the theory that MBTCl does not participate in the induction of the developmental toxicity of butyltins.

In Vitro *Dysmorphogenic Effects of Butyltin Compounds*

Krowke et al. (1986) evaluated the effects of TBTO on limb differentiation. In the organ culture system using mouse limb buds, TBTO interfered with morphogenetic differentiation at a concentration of 0.03 µg/mL. TBTO affected the differentiation of the paw skeleton and the development of the scapula. They concluded that the effects of TBTO on mouse limb differentiation should be interpreted as a cytotoxic effect rather than a specific dysmorphogenic action. Yonemoto et al. (1993) determined the relative teratogenic potencies of TBTO, TBTCl, (3-OH) hydroxybutyl dibutyltin chloride (3-OHHDBTCl), DBTCl, and MBTCl by comparing developmental hazard estimates using rat embryo limb bud cell cultures. The organotin compounds tested, except for MBTCl, were very strong inhibitors of cell differentiation

and cell proliferation. Fifty percent inhibition concentration for cell proliferation (IP50) and for cell differentiation (ID50), and the ratio of the former to
the later (P/D ratio) of each compound was determined. Among TBTO,
TBTCl, and its metabolites (i.e., 3-OHHDBTCl, DBTCl, and MBTCl), DBTCl
showed the lowest ID50 and the highest P/D ratio, therefore the teratogenic
potential of DBTCl was considered to be the highest. They noted that the
proximate toxicant of DBT teratogenicity is DBT itself, TBT is rather embryolethal than teratogenic. These findings support the results of *in vivo* developmental toxicity studies on butyltins. The embryotoxicity and dysmorphogenic potential of DBTCl were determined for gestation day 8.5 rat embryos,
which are highly susceptible to the teratogenic effects of DBTCl when administered to pregnant rats. Markedly decreased incidences in embryos with
well-developed vascularization in the body and yolk sac, yolk sac diameter,
crown-rump length, and number of somite pairs were found at 30 ng/mL
(Ema et al. 1995c). A concentration-dependent decrease in the morphological
score and increase in incidence of embryos with anomalies were noted, and
the differences were significant for embryos exposed to DBTCl at concentrations of 10 and 30 ng/mL. Open anterior neuropore and craniofacial abnormalities were predominantly observed. These results indicate that DBTCl
exerts dysmorphogenic effects on postimplantation embryos *in vitro*. Noda
et al. (1994) reported that DBT was detected in rat maternal blood at 100 ng/
g, and in embryos at 720 ng/g, at 24 hours after gavage administration of
DBTA at 22 mg/kg, teratogenic dose, on day 8 of pregnancy. Their results
show that DBT is transferred to embryos, and embryonic levels of DBT
exceed those in maternal blood, suggesting that embryos may be able to
accumulate DBT. The dysmorphogenic concentrations of DBTCl in embryos
cultured from gestation day 8.5 were well within the range of levels detected
in maternal blood after the administration of a teratogenic dose of DBT.
These findings indicate that teratogenic effects of DBTCl may be due to a
direct interference with embryos. The toxic effects of DTBCl were examined
in rat embryos during three different stages of organogenesis (i.e., the prim-
itive streak, neural fold, and early forelimb bud stages), using the rat whole
embryo culture system (Ema et al. 1996b). Rat embryos were explanted on
gestation day 8.5, 9.5, or 11.5 and cultured. Dysmorphogenesis in embryos
cultured from gestation day 8.5, 9.5, or 11.5 was observed at concentrations
of 10 ng/mL and higher, 50 ng/mL and higher, and 300 ng/mL, respectively.
Incomplete turning and craniofacial defects in embryos cultured from gestation day 8.5 and day 9.5, and defects of the forelimb buds and tail in
embryos cultured from gestation day 11.5, were frequently observed. These
results show that *in vitro* exposure to DBTCl interferes with normal development of embryos during three different stages of organogenesis and that
the susceptibility to the embryotoxicity, including dysmorphogenic potential, of DBTCl varies with developmental stage. These findings suggest that
the phase specificity for the *in vivo* teratogenesis of DBTCl given to pregnant
rats may be attributable to a decline in the susceptibility of embryos to the
dysmorphogenesis of DBTCl with advancing development.

Summary of Developmental Toxicity of Butyltin Compounds

Maternal exposure during pregnancy to TBTs, such as TBTO, TBTA, and TBTCl, caused embryonic/fetal deaths and suppression of fetal growth at maternal toxic doses. At severely maternal toxic doses of TBTs, cleft palate was produced in fetuses. Behavioral changes were also reported in postnatal offspring of rats that received TBTs during pregnancy at doses that did not cause overt maternal toxicity. Significant effects on growth profiles in male and female offspring, and decreased liver weights in female offspring were noted after administration of TBTCl by gavage from day 8 of pregnancy until adulthood even at 0.025 mg/kg. Many reports showed that DBT is teratogenic when administered during organogenesis. DBT may increase the incidence of fetal malformations at marginal doses that induced maternal toxicity. Developing embryos were not susceptible to teratogenicity of DBTCl on day 6; day 7 was the earliest susceptible period, day 8 was the most susceptible period, and day 9 was no longer a period of susceptibility to the teratogenicity of DBTCl. There were differences in the manifestation and degree of developmental toxicity among TeBT, TBT, DBT, and MBT. The developmental toxicity studies on butyltins suggest that the teratogenicity of DBT is different from those of TeBT, TBT, and MBT in its mode of action, because the susceptible period for teratogenicity and types of malformations induced by DBT are different from those induced by tetra-, tri-, and mono-substituted organotins. DBTCl exerts dysmorphogenic effects on postimplantation embryos *in vitro*. The dysmorphogenic concentrations of DBTCl in embryos cultured were well within the range of levels detected in maternal blood after the administration of a teratogenic dose of DBT. The phase specificity for the *in vivo* teratogenesis of DBTCl may be attributable to a decline in the susceptibility of embryos to the dysmorphogenesis of DBTCl with advancing development. The findings of *in vivo* and *in vitro* studies suggest that DBT itself is a causative agent in DBT teratogenesis.

Developmental Toxicity of Miscellaneous Organotin Compounds

Table 3.5 presents the developmental toxicity studies on miscellaneous organotin compounds. Behavioral effects were determined in offspring of female SD rats given trimethyltin chloride (TMTCl) in drinking water at a concentration of 0.2, 0.8, or 1.7 mg/L, or monomethyltin trichloride (MMTCl) in drinking water at a concentration of 24.3, 80.9, or 243 mg/L from 12 days before mating, to day 21 of lactation, throughout the mating and pregnancy period (Noland et al. 1982). Only male pups were tested. Learning deficiency was detected in organotin-treated pups. Pups from dams exposed to TMTCl at 1.7 mg/L or MMTCl at 243 mg/L displayed an increased acquisition time in a runway learning test on PND 11. A higher escape time in a swim escape test on PND 21 was also observed in male pups exposed to prenatal MMTCl at 24 and 243 mg/L. In this study, there was no difference between the weights of control and experimental animals in suckling pups and their

Table 3.5 Developmental Toxicity of Miscellaneous Organotin Compounds

Compounds	Animals	Dose	Days of Administration	Route	Reproductive and Developmental Effects	Author(s)
TMTCl	SD rat	1.7 mg/L	14 days before mating to lactation day 21	Drinking water	Learning deficiency in male pups	Noland et al. (1982)
MMTCl	SD rat	243 mg/L	As above	As above	As above	
TMTCl	SD rat	5–9 mg/kg	Day 7, 12, or 17 of pregnancy	ip	Decreased postnatal wt. gain, decreased no. of pups, degenerative changes in hippocampus	Paule et al. (1986)
TMTCl	THA rat	5–7 mg/kg	Day 12 of pregnancy	ip	Disruption of learning acquisition	Miyake et al. (1989)
THTCl	SD rat	5 mg/kg	Day 6–20 of pregnancy	Gavage	Increased spontaneous activity, increased d-amphetamine-stimulate rearing	Gårdlund et al. (1991)
DMTCl	Wistar rat	15–20 mg/kg	Days 7–17 of pregnancy	Gavage	Decreased fetal wt., cleft palate	Noda (2001)
		40 mg/kg	Days 7–9 or 13–15 of pregnancy	Gavage	Skeletal variations	
Octyltin stabilizer ZK 30.434 (80% DOTTG and 20% MOTTG)	Han:NMRI mouse	20–100 mg/kg	Days 5–16 of pregnancy	Gavage	Postimplantation loss, decreased fetal wt, bent forelimb, cleft palate, exencephaly, skeletal malformations and variations	Faqi et al. (2001)

dams. Postnatal growth and neuronal alterations were evaluated in pups of SD rats intraperitoneally injected on either day 7, 12, or 17 of pregnancy with a single dose of TMTCl at 5, 7, or 9 mg/kg (Paule et al. 1986). Maternal body weight at term of pregnancy was lower in the TMTCl-treated groups. Prenatal TMTCl decreased pup weight at 7 mg/kg and higher. A decreased number of surviving pups was found only in the group treated TMTCl at 9 mg/kg on day 17 of pregnancy. Generative changes in the hippocampus were more frequently noted in pups exposed to TMTCl on day 12 or 17 than on day 7. Paule et al. (1986) concluded that prenatal exposure to TMTCl causes toxic effects in postnatal offspring, but only in the presence of maternal toxicity. Disruption of learning acquisition was reported in offspring of THA rats intraperitoneally injected with TMTCl at 5 or 7 mg/kg on day 12 of pregnancy (Miyake et al. 1989). No maternal toxicity was found at 5 mg/kg. No effects of TMTCl on body weight, survival, or physical and functional development of pups were detected. In the Sidman avoidance test, the avoidance rate of the TMTCl-treated offspring rats was lower when compared to that of the controls.

Postnatal behavioral changes in pups were determined in rats prenatally administered trihexyltin chloride (THTCl) (Gårdlund et al. 1991). Pregnant SD rats were gavaged THTCl at 5 mg/kg on days 6 to 20 of pregnancy and allowed to litter. An increase in spontaneous activity, including locomotion and total activity, and a marginally increased d-amphetamine–stimulated rearing behavior were observed in postnatal pups at 5 mg/kg. This dose level did not induce maternal toxicity.

Dimethyltin chloride (DMTCl) was given to Wistar rats by gavage at 5, 10, 15, or 20 mg/kg on days 7 to 17 of pregnancy (Noda 2001). At 20 mg/kg, severe clinical signs of toxicity, including death and marked decreases in body weight gain and food consumption in pregnant rats, and incidence of cleft palate in fetuses were observed. Decreases in maternal thymus weight and fetal weight were found at 15 mg/kg and higher. No increase in incidence of fetal malformations was detected following administration of DMTCl on days 7 to 9, on days 10 to 12, on days 13 to 15, or on days 16 to 17 of pregnancy at 20 or 40 mg/kg. Noda (2001) concluded that DMTCl produced fetal malformations at a severely maternal toxic dose.

The octyltin stabilizer ZK 30.434, a mixture of 80% dioctyltin diisooctylthioglycolate and 20% monooctyltin triisooctylthioglycolate (DOTTG/MOTTG) was gavaged to Han:NMRI mice at 20, 30, 45, 67, or 100 mg/kg on days 5 to 16 of pregnancy (Faqi et al. 2001). One death at 100 mg/kg and a decreased thymus weight at 45 and 100 mg/kg were observed in dams. An increase in resorptions and low fetal weight were found at 67 mg/kg and higher. An increase in number of external and skeletal anomalies, such as forelimb bent, cleft palate, exencephaly, clavicula bent, femur bent, and fused ribs, were observed at the highest dose. Incidences of cervical and lumbar ribs were increased at 20 mg/kg and higher. These results indicate that DOTTG/MOTTG is developmentally toxic in mice.

Summary of Developmental Toxicity of Miscellaneous Organotin Compounds

Prenatal and/or postnatal exposure to TMTCl possesses developmental neu-rotoxic effects in postnatal rat offspring, even at doses that induced no maternal toxicity. The learning deficiency induced by prenatal TMTCl may be due to hippocampal lesions. Prenatal treatment of maternal toxic doses of TMTCl adversely affected survival and growth of offspring. Prenatal treatment of THTCl is also reported to induce behavioral changes in post-natal offspring. An increased number of cleft palates were observed in fetuses of rats given DMTCl during organogenesis at a severely maternal toxic dose. A mixture of DOTTG and MOTTG is developmentally toxic and produces fetal malformations in mice.

Conclusions

Many studies on toxic effects of phenyltins and butyltins in aquatic organ-isms have been conducted. TBT or TPT causes the imposition of male sex organs (termed *imposex*) on female mud snails above the concentration of about 1 ng/L (Sn) in seawater, but DBT or MPT does not induce imposex. The intensity is characterized by a classification system based on the VDS index, and in advanced phases of imposex and sterilization with gross mor-phological changes would be irreversible. The biochemical mechanism stud-ies suggested that the induction of either neurotropic hormone or androgen titers would lead to imposex induction at extremely low doses of TBT. Also TBT or TPT exposure in early life stages of fish causes altered embryonic development, impaired morphological development, and delayed or inhib-ited hatching, and induces reduced fecundity and sperm counts as repro-ductive effects. Such reproductive and developmental defects were also found in other species. The impaired reproduction and subsequent popula-tion decline in a variety of aquatic organisms by organotins are important issues in the aquatic ecosystem.

Many reports on reproductive and developmental toxic effects of phe-nyltins and butyltins in experimental animals have been published. While TPTs caused decreases in male fertility due to degenerative changes in tes-ticular tissue, the female reproductive failure induced by TPTs is more prom-inent and the harmful effects of TPTs on the ovaries were presented after five days of treatment. TPTCl during early pregnancy caused implantation failure. Implantation failure due to TPTCl might be mediated by the sup-pression of uterine decidualization and correlated with the reduction in serum progesterone levels. These findings were also shown in rats given DPT, a major metabolite of TPT. Maternal exposure to TPTs during organo-genesis caused embryonic/fetal death and suppression of fetal growth at maternal toxic doses. TPTs did not induce an increased number of fetal malformations, even at doses that produced overt maternal toxicity. Behav-ioral changes were reported in postnatal offspring of maternal rats that received TPTs during pregnancy at doses that did not cause overt maternal toxicity. In a rat two-generation reproductive toxicity study, TBTCl at relatively low doses affected male and female reproductive systems, including decreased weights of the male reproductive organs, decreased counts of spermatids and sperms, decrease in serum estradiol levels, delayed vaginal opening, impaired estrous cyclicity, and increased female AGD. TBTCl and DBTCl during early pregnancy caused implantation failure in rats. Implantation failure due to TBTCl and DBTCl, at lower doses than TBTCl, may be mediated via the suppression of uterine decidualization and correlated with the reduction in serum progesterone levels. Administration of MBTCl during early pregnancy did not cause pre- or postimplantation loss. Maternal exposure during pregnancy to TBTs caused embryonic/fetal deaths, suppression of fetal growth, and cleft palate at maternal toxic doses. Significant effects on growth profiles and decreased liver weights were reported in offspring of rats given TBTCl by gavage, even at 0.025 mg/kg from day 8 of pregnancy until adulthood. Behavioral changes were also shown in postnatal offspring of rats that received TBTs during pregnancy at doses that did not cause overt maternal toxicity. Many reports demonstrated that DBT derivatives with different anions, such as dichloride, diacetate, maleate, dilaurate, and oxide, are teratogenic when administered during organogenesis in rats. Rat embryos are the most susceptible to teratogenic effects of DBT on day 8 of pregnancy after maternal exposure. The developmental toxicity studies on butyltins suggest that the teratogenic effects of DBT are different from those of TeBT, TBT, and MBT in its mode of action. DBTCl exerts dysmorphogenic effects on postimplantation embryos *in vitro*. The phase specificity for the in vivo teratogenic effects of DBTCl may be attributable to a decline in the susceptibility of embryos to the dysmorphogenesis of DBTCl with advancing development. The findings of *in vivo* and *in vitro* studies suggest that DBT itself is a causative agent in DBT teratogenesis. Because the teratogenicity of DTB has been reported in a single species, studies in additional species would be of great value in evaluating developmental toxicity of DBT. As for miscellaneous organotin compounds, several reports on developmental toxicity are published. Prenatal and/or postnatal exposure to TMTCl or THTCl caused behavioral changes in postnatal rat offspring. Behavioral changes in postnatal pups of rats given organotin prenatally and/or postnatally may be a sensitive parameter for reproductive and developmental toxicity. A mixture of DOTTG and MOTTG is developmentally toxic and produces fetal malformations in mice. An increased number of cleft palates was reported in fetuses of rats given DMTCl during organogenesis at severely maternal toxic dose.

References

Adeeko, A., Li, D., Forsyth, D.S., Casey, V., Cooke, G.M., Barthelemy, J., Cyr, D.G., Trasler, J.M., Robaire, B., and Hales, B.F. (2003) Effects of in utero tributyltin chloride exposure in the rat on pregnancy outcome, *Toxicological Sciences*, 74, 407–15.

Baroncelli, S., Karrer, D., and Turillazzi, P.G. (1990) Embryotoxic evaluation of bis(tri-*n*-butyltin)oxide (TBTO) in mice, *Toxicology Letters*, 50, 257–62.

Baroncelli, S., Karrer, D., and Turillazzi, P.G. (1995) Oral bis(tri-*n*-butyltin) oxide in pregnant mice. I. Potential influence of maternal behavior on postnatal mortality, *Journal of Toxicology and Environmental Health*, 46, 355–67.

Belfoid, A.C., Purperhart, M., and Ariese, F. (2000) Organotin levels in seafood, *Marine Pollution Bulletin*, 40, 226–32.

Bettin, C., Oehlmann, J., and Stroben, E. (1996) YBY-induced imposex in marine neogastropods is mediated by an increasing androgen level, *Helgoländer Meeresunters*, 50, 299–317.

Boyer, I.J. (1989) Toxicity of dibutyltin, tributyltin and other organotin compounds to humans and to experimental animals, *Toxicology*, 55, 253–98.

Bryan, G.W., Gibbs, P.E., Hummerstone, L.G., and Burt, G.R. (1986) The decline of the gastropod *Nucella lapillus* around South-West England: evidence for the effect of tributyltin from antifouling paints, *Journal of the Marine Biological Association of the United Kingdom*, 66, 611–40.

Bryan, G.W., Gibbs, P.E., Burt, G.R., and Hummerstone, L.G. (1987) The effect of tributyltin (TBT) accumulation on adult dog-whelks, *Nucella lapillus*: long-term field and laboratory experiments, *Journal of the Marine Biological Association of the United Kingdom*, 67, 525–44.

Bryan, G.W., Gibbs, P.E., Hummerstone, L.G., and Burt. G.R. (1988) Comparison of the effectiveness of tri-*n*-butyltin chloride and five other organotin compounds in promoting the development of imposex in the dogwhelk *Nucella lapillus*, *Journal of the Marine Biological Association of the United Kingdom*, 68, 733–44.

Bryan, G.W., Gibbs, P.E., Hummerstone, L.G., and Burt, G.R. (1989) Uptake and transformation of 14-C labelled tributyltin chloride by the dog-whelk, *Nucella lapillus*: Importance of absorption from the diet, *Marine Environmental Research*, 28, 241–5.

Bui, Q.Q., Tran, M.B., and West, W.L. (1986) A comparative study of the reproductive effects of methadone and benzo[*a*]pyrene in the pregnant and pseudopregnant rat, *Toxicology*, 42, 195–204.

Chernoff, N., Setzer, R.W., Miller, D.B., Rosen, M.B., and Rogers, J.M. (1990) Effects of chemically induced maternal toxicity on prenatal development in the rat, *Teratology*, 42, 651–8.

Clark, R.L., Anderson, C.A., Prahalada, S., Robertson, R.T., Lochry, E.A., Leonard, Y.M., Stevens, J.L., and Hoberman, A.M. (1993) Critical developmental periods for effects on male rat genitalia induced by finasteride, a 5 alpha-reductase inhibitor, *Toxicology and Applied Pharmacology*, 119, 34–40.

Colborn, T., vom Saal, F., and Soto, A. (1993) Developmental effects of endocrine-disrupting chemicals in wildlife and humans, *Environmental Health Perspectives*, 101, 378–84.

Cooke, G.M., Tryphonas, H., Pulido, O., Caldwell, D., Bondy, G.S., and Forsyth, D. (2004) Oral (gavage), in utero and postnatal exposure of Sprague-Dawley rats to low doses of tributyltin chloride. Part I: toxicology, histopathology and clinical chemistry, *Food and Chemical Toxicology*, 42, 211–20.

Crofton, K.M., Dean, K.F., Boncek, V.M., Rosen, M.B., Sheets, L.P., Chernoff, N., and Reiter, L.W. (1989) Prenatal or postnatal exposure to bis(tri-*n*-butyltin)oxide in the rat: postnatal evaluation of teratology and behavior, *Toxicology and Applied Pharmacology*, 97, 113–23.

Cummings, A.M. (1990) Toxicological mechanisms of implantation failure, *Fundamental and Applied Toxicology*, 15, 571–9.

Davies, J.M., Bailey, S.K., and Moore, D.C. (1987) Tributyltin in Scottish sea lochs, as indicated by degree of imposex in the dogwhelk, *Nucella lapillus* (L.), *Marine Pollution Bulletin*, 18, 400–4.

Davis, A., Barale, R., Brun, G., Forster, R., Güunther, T., Hautefeuille, H., van der Heijden, C.A., Knaap, A.G.A.C., Krowke, R., Kuroki, T., Loprieno, N., Malaveille, C., Merker, H.J., Monaco, M., Mosesso, P., Nuebert, D., Norppa, H., Sorsa, M., Vogel, E., Voogd, C.E., Umeda, M., and Bartsch, H. (1987) Evaluation of the genetic and embryotoxic effects of bis(tri-*n*-butyltin)oxide (TBTO), a broad-spectrum pesticide, in multiple in vivo and in vitro short-term tests, *Mutation Research*, 188, 65–95.

De Feo, V.J. (1963) Temporal aspect of uterine sensitivity in the pseudopregnant or pregnant rat, *Endocrinology*, 72, 305–16.

Doering, D.D., Stechelbroeck, S., Doering, T., and Klingm, D. (2002) Effects of butyltins on human 5-reductase type 1 and type 2 activity, *Steroids*, 67, 859–67.

Eisler, R. (2000) Tin, in *Handbook of Chemical Risk Assessment: Health Hazards to Humans, Plants, and Animals, Volume 1 (Metals)*, Boca Raton, FL, Lewis Publishers, 551–603.

Ema, M. and Harazono, A. (2000) Adverse effects of dibutyltin dichloride on initiation and maintenance of rat pregnancy, *Reproductive Toxicology*, 14, 451–6.

Ema, M. and Harazono, A. (2001) Toxic effects of butyltin trichloride during early pregnancy in rats, *Toxicology Letters*, 125, 99–106.

Ema, M. and Miyawaki, E. (2001) Role of progesterone on suppression of uterine decidualization and implantation failure induced by triphenyltin chloride in rats, *Congenital Anomalies*, 41, 106–11.

Ema, M. and Miyawaki, E. (2002) Suppression of uterine decidualization correlated with reduction in serum progesterone levels as a cause of preimplantation embryonic loss induced by diphenyltin in rats, *Reproductive Toxicology*, 16, 309–17.

Ema, M., Itami, T., and Kawasaki, H. (1991) Teratogenicity of di-*n*-butyltin dichloride in rats, *Toxicology Letters*, 58, 347–56.

Ema, M., Itami, T., and Kawasaki, H. (1992) Susceptible period for the teratogenicity of di-*n*-butyltin dichloride in rats, *Toxicology*, 73, 81–92.

Ema, M., Kurosaka, R., Amano, H., and Ogawa, Y. (1995a) Further evaluation of the developmental toxicity of tributyltin chloride in rats, *Toxicology*, 96, 195–201.

Ema, M., Kurosaka, R., Amano, H., and Ogawa, Y. (1995b) Comparative developmental toxicity of butyltin trichloride, dibutyltin dichloride and tributyltin chloride in rats, *Journal of Applied Toxicology*, 15, 297–302.

Ema, M., Iwase, T., Iwase, Y., and Ogawa, Y. (1995c) Dysmorphogenic effects of di-*n*-butyltin dichloride in cultured rat embryos, *Toxicology in Vitro*, 9, 703–9.

Ema, M., Kurosaka, R., Amano, H., and Ogawa, Y. (1996a) Comparative developmental toxicity of di-, tri- and tetrabutyltin compounds after administration during late organogenesis in rats, *Journal of Applied Toxicology*, 16, 71–6.

Ema, M., Iwase, T., Iwase, Y., Ohyama, N., and Ogawa, Y. (1996b) Change of embryotoxic susceptibility to di-*n*-butyltin dichloride in cultured rat embryos, *Archives of Toxicology*, 70, 742–8.

Ema, M., Miyawaki, E., Harazono, A., and Ogawa, Y. (1997a) Effects of triphenyltin chloride on implantation and pregnancy in rats, *Reproductive Toxicology*, 11, 201–6.

Ema, M., Harazono, A., Miyawaki, E., and Ogawa, Y. (1997b) Effect of the day of administration on the developmental toxicity of tributyltin chloride in rats, *Archives of Environmental Contamination and Toxicology*, 33, 90–6.

Ema, M., Miyawaki, E., Harazono, A., and Ogawa, Y. (1998) Reproductive effects of butyl benzyl phthalate in pregnant and pseudopregnant rats, *Reproductive Toxicology*, 12, 127–32.

Ema, M., Miyawaki, E., and Kawashima, K. (1999a) Suppression of uterine decidualization as a cause of implantation failure induced by triphenyltin chloride in rats, *Archives of Toxicology*, 73, 175–9.

Ema, M., Miyawaki, E., and Kawashima, K. (1999b) Adverse effects of diphenyltin dichloride on initiation and maintenance of pregnancy in rats, *Toxicology Letters*, 108, 17–25.

Ema, M., Miyawaki, E., and Kawashima, K. (1999c) Developmental toxicity of triphenyltin chloride after administration on three consecutive days during organogenesis in rats, *Bulletin of Environmental Contamination and Toxicology*, 62, 363–70.

Ema, M., Miyawaki, E., and Kawashima, K. (2000) Critical period for adverse effects on development of reproductive system in male offspring of rats given di-*n*-butyl phthalate during late pregnancy, *Toxicology Letters*, 111, 271–8.

Ema, M., Harazono, A., Hirose, A., and Kamata, E. (2003) Protective effects of progesterone on implantation failure induced by dibutyltin dichloride in rats, *Toxicology Letters*, 143, 233–8.

Epstein, S.S., Arnold, E., Andrea, J., Bass, W., and Bishop, Y. (1972) Detection of chemical mutagens by the dominant lethal assay in the mouse, *Toxicology and Applied Pharmacology*, 23, 288–325.

Evans, D.W. and Laughlin, R.B., Jr. (1984) Accumulation of bis(tributyltin) oxide by the mud crab, *Rhithropanopeus harrisii*, *Chemosphere*, 13, 213–9.

Faqi, A.S., Schweinfurth, H., and Chahoud, I. (1997) Determination of the no-effect dose of bis(tri-*n*-butyltin)oxide (TBTO) for maternal toxicity and teratogenicity in mice, *Congenital Anomalies*, 37, 251–8.

Faqi, A.S., Schweinfurth, H., and Chahoud, I. (2001) Developmental toxicity of an octyltin stabilizer in NMRI mice, *Reproductive Toxicology*, 15, 117–22.

Farr, C.H., Reinisch, K., Holson, J.F., and Neubert, D. (2001) Potential teratogenicity of di-*n*-butyltin dichloride and other dibutyltin compounds, *Teratogenesis, Carcinogenesis, and Mutagenesis*, 21, 405–15.

Fent, K. (1996) Ecotoxicology of organotin compounds, *Critical Reviews in Toxicology*, 26, 1–117.

Fent, K. and Hunn, J. (1991) Phenyltins in water, sediment, and biota of freshwater marinas, *Environmental Science and Technology*, 25, 956–63.

Fent, K. and Meier, W. (1992) Tributyltin-induced effects on early life stages of minnows *Phoxinus phoxinus*, *Archives of Environmental Contamination and Toxicology*, 22, 428–31.

Fent, K. and Meier, W. (1994) Effects of triphenyltin on fish early life stages, *Archives of Environmental Contamination and Toxicology*, 27, 224–31.

Féral, C. and LeGall, S. (1983) The influence of a pollutant factor (tributyltin) on the neuroendocrine mechanism responsible for the occurrence of a penis in the female of *Ocenebra erinacea*, in *Molluscan neuro-endocrinology*, Lever, J. and Boer, H., Eds., Amsterdam: North Holland Publishing Co., 173–5.

Fioroni, P., Oehlmann, J., and Stroben, E. (1991) The pseudohermaphroditism of prosobranchs; morphological aspects, *Zoologischer Anzeiger*, 226, 1–26.

Fish, R.H., Kimmel, E.C., and Casida, J.E. (1976) Bioorganotin chemistry: reactions of tributyltin derivatives with a cytochrome P-450 dependent monooxygenese enzyme system, *Journal of Organometallic Chemistry*, 118, 41–54.

Gaines, T.B. and Kimbrough, R.D. (1968) Toxicity of fentin hydroxide to rats, *Toxicology and Applied Pharmacology*, 12, 397–403.

Gårdlund, A.T., Archer, T., Danielsson, B., Fredriksson, A., Lindqvist, N.G., Lindstrom, H., and Luthman, J. (1991) Effects of prenatal exposure to tributyltin and trihexyltin on behaviour in rats, *Neurotoxicology and Teratology*, 13, 99–105.

Giavini, E., Prati, M., and Vismara, C. (1980) Effects of triphenyltin acetate on pregnancy in the rat, *Bulletin of Environmental Contamination and Toxicology*, 24, 936–9.

Gibbs, P.E. and Bryan, G.W. (1986) Reproductive failure in populations of the dog-welk, *Nucella lapillus*, caused by imposex induced by tributyltin from antifouling paints, *Journal of the Marine Biological Association of the United Kingdom*, 66, 767–77.

Gibbs, P.E., Bryan, G.W., Pascoe, P.L., and Burt, G.R. (1987) The use of the dog-whelk, *Nucella lapillus*, as an indicator of tributyltin (TBT) contamination, *Journal of the Marine Biological Association of the United Kingdom*, 67, 507–23.

Gibbs, P.E., Pascoe, P.L., and Burt, G.R. (1988) Sex change in the female dog-whelk, *Nucella lapillus*, induced by tributyltin from antifouling paints, *Journal of the Marine Biological Association of the United Kingdom*, 68, 715–31.

Harazono, A. and Ema, M. (2000) Suppression of decidual cell response induced by tributyltin chloride in pseudopregnant rats: as a cause of early embryonic loss, *Archives of Toxicology*, 74, 632–7.

Harazono, A. and Ema, M. (2003) Suppression of decidual cell response induced by dibutyltin dichloride in pseudopregnant rats: As a cause of early embryonic loss, *Reproductive Toxicology*, 17, 393–9

Harazono, A., Ema, M., and Kawashima, K. (1998a) Evaluation of malnutrition as a cause of tributyltin-induced pregnancy failure in rats, *Bulletin of Environmental Contamination and Toxicology*, 61, 224–30.

Harazono, A., Ema, M., and Kawashima, K. (1998b) Evaluation of early embryonic loss induced by tributyltin chloride in rats: Phase- and dose-dependent antifertility effects, *Archives of Environmental Contamination and Toxicology*, 34, 94–9.

Harazono, A., Ema, M., and Ogawa, Y. (1996) Pre-implantation embryonic loss induced by tributyltin chloride in rats, *Toxicology Letters*, 89, 185–90.

Haubruge, E., Petit, F., and Gage, M.J. (2000) Reduced sperm counts in guppies (*Poecilia reticulata*) following exposure to low levels of tributyltin and bisphenol A, *Proceedings of the Royal Society of London. Series B. Biological Sciences*, 267, 2333–7.

Hawkins, L.E. and Hutchinson, S. (1990) Physiological and morphogenetic effects of monophenyltin trichloride on *Ocenebra erinacea* (L.), *Functional Ecology*, 4, 449–54.

Holm, G., Norrgren, L., and Linden, O. (1991) Reprodictive and histophathological effects of long-term experiemental exposure to bis(tributyltin) oxide (TBTO) on the three-spined stickeback, *Gasterosteus aculeatus* Linnaeus, *Journal of Fish Biology*, 38, 373–86.

Horiguchi, T., Shiraishi, H., Shibata, Y., Morita, M., and Shimizu, M. (1996) Imposex and organotin contamination in gastropods after the regulation of organotin use in Japan, *Abstract Book of SETAC 17th Annual Meeting*, 177.

Horiguchi, T., Shiraishi, H., Shimizu, M., and Morita, M. (1997a) Imposex in sea snails, caused by organotin (tributyltin and triphenyltin) pollution in Japan: a survey, *Applied Organometallic Chemistry*, 11, 451–5.

Horiguchi, T., Shiraishi, H., Shimizu, M., and Morita, M. (1997b) Effects of triphenyltin chloride and five other organotin compounds on the development of imposex in the rock shell thais clavigera, *Environmental Pollution*, 95, 85–91.

Horiguchi, T., Kojima, M., Kaya, M., Matsuo, T., Shiraishi, H., Morita, M., and Adachi, Y. (2002) Tributyltin and triphenyltin induce spermatogenesis in ovary of female abalone, Haliotis gigantea, *Marine Environmental Research*, 54, 679–84.

International Programme on Chemical Safety (IPCS) (1999a) *Concise International Chemical Assessment Document, No. 14 Tributyltin Oxide*, IPCS, Geneva, World Health Organization.

International Programme on Chemical Safety (1999b) *Concise International Chemical Assessment Document, No. 13 Triphenyltin Compounds*, Geneva, World Health Organization.

Ishizaka, T., Suzuki, T., and Saito, Y. (1989) Metabolism of dibutyltin dichloride in male rats, *Journal of Agricultural and Food Chemistry*, 37, 1096–101.

Itami, T., Ema, M., Amano, H., Murai, T., and Kawasaki, H. (1990) Teratogenic evaluation of tributyltin chloride in rats following oral exposure, *Drug and Chemical Toxicology*, 13, 283–95.

Iwai, H., Wada, O., and Arakawa, Y. (1981) Determination of tri-, di-, and monobutyltin and inorganic tin in biological materials and some aspects of their metabolism in rats, *Journal of Analytical Toxicology*, 5, 300–6.

Japan Environment Agency (1998) *Strategic Programs on Environmental Endocrine Disruptors '98*, Tokyo, Environmental Health Department.

Kamrin, M.A., Carney, E.W., Chou, K., Cummings, A., Dostal, L.A., Harris, C., Henck, J.W., Loch-Caruso, R., and Miller, R.K. (1994) Female reproductive and developmental toxicology: overview and current approaches, *Toxicology Letters*, 74, 99–119.

Kannan, K., Tanabe, S., and Tatsukawa, R. (1995) Phenyltin residues in horseshoe crabs, *Tachypleus tridentatus* from Japanese coastal waters, *Chemosphere*, 30, 925–32.

Kannan, K., Corsolini, S., Focardi, S., Tanabe, S., and Tatsukawa, R. (1996) Accumulation pattern of butyltin compounds in dolphin, tuna, and shark collected from Italian coastal waters, *Archives of Environmental Contamination and Toxicology*, 31, 19–23.

Karrer, D., Baroncelli, S., and Turillazzi, P.G. (1995) Oral bis(tri-*n*-butyltin) oxide in pregnant mice. II. Alterations in hematological parameters, *Journal of Toxicology and Environmental Health*, 46, 369–77.

Kenaga, E.E. (1965) Triphenyl tin compounds as insect reproduction inhibitors, *Journal of Economic Entomology*, 58, 4–8,

Kime, D.E., Huyskens, G., McAllister, B.G., Ruranguwa, E., Skorkowski, G., and Ollevia, F. (2001) Tributyltin disrupts the vertebrate reproductive system at muliple sites, *Abstracts 11th Annual Meeting of SETAC Europe*, Madrid, May 6–10, 195.

Kimmel, E.C., Fish, R.H., and Casida, J.E. (1977) Bioorganotin chemistry. Metabolism of organotin compounds in microsomal monooxygenase system and in mammals, *Journal of Agricultural and Food Chemistry*, 25, 1–9.

Krowke, R., Bluth, U., and Neubert, D. (1986) In vitro studies on the embryotoxic potential of (bis[tri-*n*-butyltin])oxide in a limb bud organ culture system, *Archives of Toxicology*, 58, 125–9.

Kumasaka, K., Miyazawa, M., Fujimaki, T., Tao, H., Ramaswamy, B.R., Nakazawa, H., Makino, T., and Satoh, S. (2002) Toxicity of the tributyltin compound on the testis in premature mice, *Journal of Reproduction and Development*, 48, 591–7.

Lau, M.M. (1991) Tributyltin antifoulings: a threat to the Hong Kong marine environment, *Archives of Environmental Contamination and Toxicology*, 20, 299–304.

Laughlin, R.B., Jr., French, W., and Guard, H.E. (1986) Accumulation of bis(tributyltin) oxide by the marine mussel *Mytilus edulis*, *Environmental Science and Technology*, 20, 884–90.

Lehotzky, K., Szeberenyi, J.M., Gonda, Z., Horkay, F., and Kiss, A. (1982) Effects of prenatal triphenyl-tin exposure on the development of behavior and conditioned learning in rat pups, *Neurobehavioral Toxicology and Teratology*, 4, 247–50.

Maguire, R.J. (1991) Aquatic environmental aspects of non-pesticidal organotin compounds, *Water Pollution Research Journal of Canada*, 26, 243–360.

Manning, C.S., Lytle, T.F., Walker, W.W., and Lytle, J.S. (1999) Life-cycle toxicity of bis(tributyltin) oxide to the sheepshead minnow (*Cyprinodon variegatus*), *Archives of Environmental Contamination and Toxicology*, 37, 258–66.

Miyake, K., Misawa, T., Aikawa, H., Yoshida, T., and Shigeta, S. (1989) The effects of prenatal trimethyltin exposure on development and learning in the rat, *Japanese Journal of Industrial Health*, 31, 363–71 (in Japanese).

Miyake, K., Misawa, T., and Shigeta, S. (1990) Toxicity of bis (tri-*n*-butyltin) oxide on learning and development in the rat, *Nippon Eiseigaku Zasshi*, 45, 926–34 (in Japanese).

Miyake, K., Misawa, T., and Shigeta, S. (1991) The effects of prenatal triphenyltin exposure on learning and development in the rat, *Nippon Eiseigaku Zasshi*, 46, 769–76 (in Japanese).

Newton, D.W. and Hays, R.L. (1968) Histological studies of ovaries in rats treated with hydroxyurea, triphenyltin acetate, and triphenyltin chloride, *Journal of Economic Entomology*, 61, 1668–9.

Nirmala, K., Oshima, Y., Lee, R., Imada, N., Honjo, T., and Kobayashi, K (1999) Transgenerational toxicity of tributyltin and its combined effects with polychlorinated biphenyls on reproductive processes in Japanese medaka (*Oryzias latipes*), *Environmental Toxicology and Chemistry*, 18, 717–21.

Noda, T. (2001) Maternal and fetal toxicity of dimetyltin in rats, *Journal of Health Science*, 47, 544–51.

Noda, T., Morita, S., and Baba, A. (1993) Teratogenic effects of various di-*n*-butyltins with different anions and butyl(3-hydroxybutyl)tin dilaurate in rats, *Toxicology*, 85, 149–60.

Noda, T., Morita, S., and Baba, A. (1994) Enhanced teratogenic activity of di-*n*-butyltin diacetate by carbon tetrachloride pretreatment in rats, *Food and Chemical Toxicology*, 32, 321–7.

Noda, T., Morita, S., Shimizu, M., Yamamoto, T., and Yamada, A. (1988) Safety evaluation of chemicals for use in house-hold products (VIII)—teratological studies on dibutyltin diacetate in rats, *Annual Report of Osaka City Institute of Public Health and Environ Sciences*, 50, 66–75 (in Japanese).

Noda, T., Morita, S., Yamano, T., Shimizu, M., and Yamada, A. (1991a) Effects of triphenyltin acetate on pregnancy in rats by oral administration, *Toxicology Letters*, 56, 207–12.

Noda, T., Morita, S., Yamano, T., Shimizu, M., Nakamura, T., Saitoh, M., and Yamada, A. (1991b) Teratogenicity study of tri-*n*-butyltin acetate in rats by oral administration, *Toxicology Letters*, 55, 109–15.

Noda, T., Yamano, T., Shimizu, M., Saitoh, M., Nakamura, T., Yamada, A., and Morita, S. (1992a) Comparative teratogenicity of di-*n*-butyltin diacetate with *n*-butyltin trichloride in rats, *Archives of Environmental Contamination and Toxicology*, 23, 216–22.

Noda, T., Nakamura, T., Shimizu, M., Yamano, T., and Morita, S. (1992b) Critical gestational day of teratogenesis by di-*n*-butyltin diacetate in rats, *Bulletin of Environmental Contamination and Toxicology*, 49, 715–22.

Noda, T., Yamano, T., and Shimizu, M. (2001) Effects of maternal age on teratogenicity of di-*n*-butyltin diacetate in rats, *Toxicology*, 167, 181–9.

Noland, E.A., Taylor, D.H., and Bull, R.J. (1982) Monomethyl-and trimethyltin compounds induce learning deficiencies in young rats, *Neurobehavioral Toxicology and Teratology*, 4, 539–44.

Oberdörster, E. and McClellan-Green, P. (2002) Mechanisms of imposex induction in the mud snail, *Ilyanassa obsoleta*: TBT as a neurotoxin and aromatase inhibitor, *Marine Environmental Research*, 54, 715–8.

Oehlmann, J., Stroben, E., and Fioroni, P. (1991) The morphlogical expression of imposex in *Nucella lapillus* (Linnaeus) (Gastropoda: Muricidae), *Journal of Molluscan Studies*, 57, 375–90.

Oehlmann, J., Stroben, E., Fioroni, P., and Markert, B. (1996) Tributyltin (TBT) effects on *Ocinebrina aciculata* (Gastropoda: Muricidae): imposex development, sterilization, sex change and population decline, *The Science of the Total Environment*, 188, 205–23.

Ogata, R., Omura, M., Shimasaki, Y., Kubo, K., Oshima, Y., Aou, S., and Inoue, N. (2001) Two-generation reproductive toxicity study of tributyltin chloride in female rats, *Journal of Toxicology and Environmental Health A*, 63, 127–44.

Ohhira, S. and Matsui, H. (1993a) Gas chromatographic determination of inorganic tin in rat urine after a single oral administration of stannous chloride and mono-, di-, and triphenyltin chloride, *Journal of Chromatography*, 622, 173–8.

Ohhira, S. and Matsui, H. (1993b) Metabolism of diphenyltin compound in rat liver after a single oral administration of diphenyltin dichloride, *Journal of Agricultural and Food Chemistry*, 41, 607–9.

Omura, M., Ogata, R., Kubo, K., Shimasaki, Y., Aou, S., Oshima, Y., Tanaka, A., Hirata, M., Makita, Y., and Inoue, N. (2001) Two-generation reproductive toxicity study of tributyltin chloride in male rats, *Toxicological Sciences*, 64, 224–32.

Pate, B.D. and Hays, R.L. (1968) Histological studies of testes in rats treated with certain insect chemosterilants, *Journal of Economic Entomology*, 61, 32–4.

Paule, M.G., Reuhl, K., Chen, J.J., Ali, S.F., and Slikker, W., Jr. (1986) Developmental toxicology of trimethyltin in the rat, *Toxicology and Applied Pharmacology*, 84, 412–7.

Piver, W.T. (1973) Organotin compounds: industrial applications and biological investigation, *Environmental Health Perspectives*, 4, 61–79.

Quevauviller, P.H., Bruchet, A., and Donard, O.F.X. (1991) Leaching of organotin compounds from poly(vinyl chloride) (PVC) material, *Applied Organometallic Chemistry*, 5, 125–9.

Sanderson, J.T., Boerma, J., Lansbergen, G.W.A., and van den Berg, M. (2002) Induction and inhibition of aromatase (CYP19) activity by various classes of pesticides in H295R human adrenocortical carcinoma cells, *Toxicology and Applied Pharmacology*, 182, 44–54.

Sasaki, K., Ishizaka, T., Suzuki, T., and Saito, Y. (1988) Determination of tri-*n*-butyltin and di-*n*-butyltin compounds in fish by gas chromatography with flame photometric detection, *Journal of the Association of Official Analytical Chemists*, 71, 360–3.

Scadding, S.R. (1990) Effects of tributyltin oxide on the skeletal structures of developing and regenerating limbs of the axolotl larvae, *Ambystoma mexicanum*, *Bulletin of Environmental Contamination and Toxicology*, 45, 574–81.

Schardein, J. (2000) Hormones and hormonal antagonists, in *Chemically Induced Birth Defects*, 3rd ed., revised and expanded, Marcel Dekker, New York.

Short, J.W. and Thrower, F.P. (1986) Accumulation of butyltins in muscle tissue of chinook salmon reared in sea pens treated with tri-*n*-butyltin, *Marine Pollution Bulletin*, 17, 542–5.

Smith, B.S. (1981a). Male characteristics on female mud snails caused by antifouling paints, *Journal of Applied Toxicology*, 1, 22–5.

Smith, B.S. (1981b). Tributyltin compounds induce male characteristics on female mud snails *Nassarius oboletus* = *Ilyanasa oboleta*, *Journal of Applied Toxicology*, 1, 141–4.

Snoeij, N.J., Penninks, A.H., and Seinen, W. (1987) Biological activity of organotin compounds—an overview, *Environmental Research*, 44, 335–53.

Snow, R.L. and Hays, R.L. (1983) Phasic distribution of seminiferous tubules in rats treated with triphenyltin compounds, *Bulletin of Environmental Contamination and Toxicology*, 31, 658–65.

Spencer, F. and Sing, L.T. (1982) Reproductive responses to rotenone during decidualized pseudogestation and gestation in rats, *Bulletin of Environmental Contamination and Toxicology*, 28, 360–8.

Spooner, N., Gibbs, P.E., Bryan, G.W., and Goad, L.J. (1991) The effect of tributyltin upon steroid titers in the female dogwhelk, *Nucella lapillus*, and the development of imposex, *Marine Environmental Research*, 32, 37–49.

Strmac, M. and Braunbeck, T. (1999) Effects of triphenyltin acetate on survival, hatching success, and liver ultrastructure of early life stages of zebrafish (*Danio rerio*), *Ecotoxicology and Environmental Safety*, 44, 25–39.

Stroben, E., Oehlmann, J., and Bettin, C. (1991) TBT-induced imposex and role of steroids in marine snails, in *Proceedings Tenth World Meeting of the Organotin Environmental Programme Association*, Berlin, September, 68–73.

Suzuki, T., Matsuda, R., and Saito, Y. (1992) Molecular species of tri-*n*-butyltin compounds in marine products, *Journal of Agricultural and Food Chemistry*, 40, 1437–43.

Takahashi, T., Araki, A., Nomura, Y., Koga M., and Arizono, K. (2000) The occurrence of dual-gender imposex in Japanese freshwater crab, *Journal of Health Science*, 46, 376–9.

Toyoda, M., Sakai, H., Kobayashi, Y., Komatsu, M., Hoshino Y., Horie, M., Saeki, M., Hasegawa, Y., Tsuji, M., Kojima, M., Toyomura, K., Kumano, M., and Tanimura, A. (2000) Daily dietary intake of tributyltin, dibutyltin, triphenyltin and diphenyltin compounds according to a total diet study in Japanese population, *Shokuhin Eiseigaku Zasshi*, 41, 280–6 (in Japanese).

Tryphonas, H., Cooke, G.M., Caldwell, D., Bondy, G., Parenteau, M., Hayward, S., and Pulido, O. (2004) Oral (gavage), in utero and postnatal exposure of Sprague-Dawley rats to low doses of tributyltin chloride. Part II: Effects on the immune system, *Food and Chemical Toxicology*, 42, 221–235.

Tsuda, T., Inoue, T., Kojima, M., and Aoki, S. (1995) Daily intakes of tributyltin and triphenyltin comounds from meals, *Journal of AOAC International*, 78, 941–3.

Tsuda, T., Nakanishi, H., Aoki, S., and Takebayashi, J. (1987) Bioconcentration and metabolism of phenyltin chlorides in carp, *Water Research*, 21, 949–53.

Ueno, S., Susa, N., Furukawa, Y., Komatsu, Y., Koyama, S., and Suzuki, T. (1999) Butyltin and phenyltin compounds in some marine fishery products on the Japanese market, *Archives of Environmental Health*, 54, 20–5.

Waldock, M.J. and Thain, J.E. (1983) Shell thickening in Crassostrea gigas: organotin antifouling or sediment induced? *Marine Pollution Bulletin*, 14, 411–5.

Weis, J.S. and Kim, K. (1988) Tributyltin is a teratogen in producing deformities in limbs of the fiddler crab, *Uca pugilator*, *Archives of Environmental Contamination and Toxicology*, 17, 583–7.

Winek, C.L., Marks, M.J., Jr., Shanor, S.P., and Davis, E.R. (1978) Acute and subacute toxicology and safety evaluation of triphenyl tin hydroxide (Vancide KS), *Clinical Toxicology*, 13, 281–96.

Winship, K.A. (1988) Toxicity of tin and its compounds, *Adverse Drug Reactions and Acute Poisoning Reviews*, 7, 19–38.

World Health Organization (1980) Tin and Organotin Compounds: A Preliminary Review. *Environmental Health Criteria 15*, World Health Organization, Geneva.

World Health Organization (1992) Fentin, in *Pesticide Residues in Food 1991: Evaluations Part II Toxicology*, World Health Organization, Geneva. Online: http://www.inchem.org/documents/jmpr/jmpmono/v91pr11.htm (accessed 16 June 2004).

Yamabe, Y., Hoshino, A., Imura, N., Suzuki, T., and Himeno, S. (2000) Enhancement of androgen-dependent transcription and cell proliferation by tributyltin and triphenyltin in human prostate cancer cells, *Toxicology and Applied Pharmacology*, 169, 177–84.

Yonemoto, J., Shiraishi, H., and Soma, Y. (1993) In vitro assessment of teratogenic potential of organotin compounds using rat embryo limb bud cell cultures, *Toxicology Letters*, 66, 183–91.

chapter 4

Adverse Effects of Aluminum, Uranium, and Vanadium on Reproduction and Intrauterine Development in Mammals

Jose L. Domingo
Laboratory of Toxicology and Environmental Health,
School of Medicine, "Rovira i Virgili" University, Reus, Spain

Contents

Toxic elements such as arsenic, cadmium, lead, and mercury have profoundly adverse effects upon reproduction and embryonic and fetal development. While a number of reviews have discussed the abundant data existing for the evidence of reproductive and developmental toxicity of these elements in mammals (Ernhart 1992; Domingo 1994, 1997; Golub 1994; Winneke et al. 1996; Golub et al. 1998), the effects of other toxic metals of potential environmental, industrial, or therapeutic concern have not been extensively reviewed. The reproductive and embryo/fetal toxicity of aluminum, uranium, and vanadium in mammals are herein discussed.

Aluminum

Until recent decades, there was little concern about toxic consequences of aluminum (Al) ingestion because it was assumed that this element was not orally bioavailable. However, it has been shown that although the gastrointestinal tract normally represents a major barrier to Al absorption, under some circumstances this barrier can be breached (Alfrey 1985, 1986). Consequently, individuals ingesting large amounts of Al compounds do absorb a definite amount of Al (Alfrey 1985). Although normally, mammals maintain very low Al concentrations in their tissues because of a combination of low intestinal uptake and rapid clearance, it is now recognized that toxicity can occur if absorption is markedly increased or renal clearance is impaired. While most foods contain small but variable amounts of Al, the exposure to Al through the diet is small compared to the quantities of Al in many antacid products, some buffered analgesics and other therapeutic preparations (Domingo 2003). In relation to this, Al-containing antacids are widely used nonprescription medications, which have been administered for many years for the treatment of various gastrointestinal disorders. During pregnancy, dyspepsia is a common complaint and antacids are widely used to reduce the dyspeptic symptoms. In most of these drugs, Al is present as Al hydroxide, which has a very low aqueous solubility. However, the consumption of high amounts of Al compounds during pregnancy can mean a potential risk of Al accumulation because of the important number of dietary constituents (ascorbate, citrate, lactate, succinate, etc.), that can enhance gastrointestinal Al absorption (Domingo et al. 1991a, b, 1994).

Until recently, information on human studies to determine whether Al ingestion could have adverse effects on the outcome of pregnancies was very scarce (Domingo 1995a; Golub and Domingo 1996, 1998). Weberg et al. (1986) reported that because it was not clear whether maternal Al could increase the Al levels in the fetus, high-dose antacids should not be consumed during pregnancy. More recently, Golding et al. (1991) carried out studies to determine whether Al sulfate accidentally added to a local water supply (Cornwall, United Kingdom) had adverse embryo/fetal effects in pregnant women. It was concluded that although a lack of major problems associated with fetal exposure to high Al doses was noted, the relatively small number of pregnancies made it impossible to say that high doses of Al sulfate would be safe during gestation. On the other hand, the literature contains relatively little information regarding either the experimental embryotoxic and teratogenic potential of Al, or the effects of gestational exposure to Al on the fetus and newborn. Data about Al-induced embryo/fetal toxicity and its potential reproductive toxicology are here reviewed with special attention to the results obtained in our laboratory.

Reproductive Effects of Aluminum

Kamboj and Kar (1964) administered a single intratesticular injection of 27.4 mg/kg of Al sulfate to rats. The testes were focally necrosed within 2 days after injection, while this dose also destroyed all the spermatozoa within 7 days after Al administration. These investigators also found that daily subcutaneous (SC) injections of 27.4 mg of Al sulfate to mice for 30 days reduced the weight of the testes and caused shrinkage of the tubules and spermatogenic arrest at the primary spermatocyte or spermatogonial stages without affecting the interstitium. In another study, Krasovskii et al. (1979) administered Al chloride to rats and guinea pigs at doses of 6, 17, and 50 mg Al/kg/day, and 3, 9, and 27 mg Al/kg/day to rabbits during 20 to 30 days (short-term exposure). Rats were also treated with 0.0025, 0.25, and 2.5 mg Al/kg/day for 6 months (chronic exposure). At the end of the short-term experiment, slight toxicity to the gonads was noted, while in the chronic exposure, the gonadotoxic effect was also weak with changes in the number of spermatozoa and in their motility, which were only observed in animals receiving the highest Al doses.

The reproductive toxicity of Al was also investigated in our laboratory. Adult male mice were treated intraperitoneally with Al nitrate at 0, 50, 100, and 200 mg/kg/day for 4 weeks before mating with untreated females. Decreased pregnancy rate was observed in the females mated with males previously exposed to 100 and 200 mg/kg/day. High-dose male mice showed significantly decreased testicular and epididymal weights, as well as significant decreases in testicular and spermatid counts and epididymal sperm counts. However, the sperm motility was unaffected, and the percentage of morphological normal spermatozoa in all Al-exposed mice was comparable to the values in control animals. Because fertility was only affected at 100 and 200 mg/kg/day, and neither inhibition of spermatogenesis nor testicular degeneration was noted at the lowest dose (50 mg/kg/day), the no observable adverse-effect level (NOAEL) of parenteral Al exposure on the mouse male reproductive system was 50 mg/kg/day (Llobet et al. 1995).

We also assessed in rats the effects on reproduction, gestation, parturition and lactation of oral Al exposure at 0, 13, 26, and 52 mg Al/kg/day given as Al nitrate nonahydrate (Domingo et al. 1987a, b). Male rats were treated orally for 60 days prior to mating with mature virgin female rats treated for 14 days prior to mating. Treatment was continued throughout mating, gestation, parturition, and weaning of the pups. One-half of the dams in each group were killed on gestation day 13, and the remaining dams were allowed to deliver and wean their offspring. Although no adverse effects on fertility or general reproductive parameters were noted, the survival ratios were higher in the control group. A dose-dependent delay in the growth of the living pups was also observed in all Al-treated groups (Domingo et al. 1987a).

The effect of long-term ingestion of Al on aggression, sexual behavior, and fertility was also investigated in male rats. Animals were exposed to Al chloride in the drinking water at 1000 ppm for 12 weeks. Adverse effects were observed. Sexual behavior was suppressed and the copulatory efficiency was reduced, while body, absolute and relative testes, and seminal vesicles weights dropped (Bataineh et al. 1998). In a recent study in Al-treated intraperitoneal (IP) mice, Guo et al. (2001) concluded that Al exerted a reversible but significant adverse effect on the steroidogenesis. In addition, serum and testicular testosterone concentrations were markedly decreased, which could be due to an increased production of nitric oxide products induced by excessive Al (Guo et al. 2001).

A summary of the above studies is presented in Table 4.1.

Embryo/Fetal Toxicity of Aluminum

It is now well established that Al may be an embryo/fetal toxin depending on the route of exposure and/or the solubility of the Al compound administered. While Al chloride was found to be embryotoxic and teratogenic when given parenterally to rats and mice (Benett et al. 1975; Wide 1984; Cranmer et al. 1986), no evidence of maternal and embryo/fetal toxicity was observed when high doses of Al hydroxide were given to pregnant rats and mice. Thus, no adverse developmental effects of Al were observed when Al hydroxide was given orally to rats at 66.5, 133, and 266 mg Al/kg/day (Gomez et al. 1990), or mice at 23, 46, and 92 mg Al/kg/day (Domingo et al. 1989a) during organogenesis. In addition, the maternal-placental Al concentrations were not significantly different between control and Al-treated rats, while Al could not be detected in the whole fetuses in any of the groups (Gomez et al. 1990). These results indicate that Al from Al hydroxide is very poorly absorbed and does not reach the fetus at levels that might mean a developmental hazard. In mice for example, the doses of Al hydroxide would be equivalent to those consumed by people of 60 kg body weight who ingest 1.4, 2.8, or 5.5 g of Al per day, respectively, which are much higher than the amounts usually ingested for peptic disorders. However, it has been recently shown that oral exposure to 200 and 400 mg/kg/day of Al hydroxide to rats on gestation days 1 to 20 produces significant changes in the tissue distribution of various essential elements (Ca, Mg, Mn, Cu, Zn, and Fe) (Bellés et al. 2001).

In contrast to these results, oral administration of Al nitrate nonahydrate (13, 26, and 62 mg Al/kg/day) to pregnant rats on gestation days 6 to 14 resulted in decreased fetal body weight and increased the incidence and types of external, visceral, and skeletal malformations and variations in all the Al-treated groups (Paternain et al. 1988). It was concluded that although

Table 4.1 Reproductive Toxicity of Aluminum: A Summary of Studies

Aluminum Compound	Species	Doses	Dosing Period	Route	Toxic Effects	Reference
Aluminum sulfate	Rats	27.4 mg/kg	Single dose	Intratesticular	Testes focally necrosed within 2 days after injection. Spermatozoa destroyed 7 days after injection	Kamboj and Kar (1964)
Aluminum chloride	Rats	6, 17, and 50 mg/kg/day	20–30 days	Oral	Slight toxicity to the gonads	Krasovskii et al. (1979)
	Guinea pigs Rabbits	3, 9, and 27 mg/kg/day	20–30 days	Oral	Slight toxicity to the gonads	
	Rats	0.0025, 0.25, and 2.5 mg/kg/day	6 months	Oral	Gonadotoxic effect: weak; changes in the number and motility of spermatozoa at 2.5 mg/kg/day	Llobet et al. (1995)
Aluminum nitrate	Male mice	50, 100, and 200 mg/kg/day	4 weeks before mating with untreated females	IP	Decreased pregnacy rate in females mated with males treated with 100 and 200 mg/kg/day. Necrosis of spermatocytes/spermatids at 100 and 200 mg/kg/day	
Aluminum chloride	Male rats	1000 ppm	12 weeks	Drinking water	Adverse effects on sexual behavior, fertility, and the reproductive system	Bataineh et al. (1998)
Aluminum chloride	Male mice	1/8 and 1/3 of LD_{50}	12–16 days	IP	Serum and testicular testosterone levels decreased; adverse effects on steroidogenesis	Guo et al. (2001)

embryolethality was not induced in rats by oral Al nitrate administration, teratogenic effects might result at Al nitrate doses as high as those administered in that study, corresponding to approximately 1/20, 1/10, and 1/5 of the acute oral LD_{50} of Al nitrate nonahydrate for adult female rats (Llobet et al. 1987). These data, together with those obtained from studies in which Al hydroxide was given orally to rats and mice (Domingo et al. 1989a; Gomez et al. 1990), show that Al compound solubility plays an essential role in the potential embryo/fetal toxicity of Al.

In recent decades, it has been demonstrated that ingestion of Al hydroxide concurrently with some common organic constituents of the diet (citrate, ascorbate, lactate, succinate, etc.) causes a marked increase in the gastrointestinal absorption of Al in healthy individuals (Slanina et al. 1986; Domingo et al. 1991a, b, 1994). The presence of Al complexing compounds in the gastrointestinal tract with gastric acid solubilizes Al cations and may thus result in the equilibrium formation of a soluble complex of Al, which by preventing reprecipitation, may result in Al absorption (Partridge et al. 1989). Taking this into account, we investigated whether the concurrent ingestion of citric, lactic, or ascorbic acid and high doses of Al hydroxide might result in developmental toxicity in mammals. The concurrent oral administration of citric acid (62 mg/kg/day) and Al hydroxide (133 mg Al/kg/day) to rats on gestation days 6 to 15 did not modify the lack of embryotoxicity and teratogenicity previously reported. However, the incidence of skeletal variations (delayed ossification of occipital and sternebrae) was significantly increased (Gomez et al. 1991). Although not significantly different, the incidence of skeletal variations also increased in fetuses of pregnant mice given oral doses of Al hydroxide (57 mg Al/kg/day) and lactic acid (570 mg/kg/day) on days 6 to 15 of gestation, in comparison with a group of animals receiving Al hydroxide (57 mg Al/kg/day) only (Colomina et al. 1992). By contrast, no signs of developmental toxicity were observed in mice when Al hydroxide (104 mg Al/kg/day) was given by gavage concurrently with high doses of ascorbic acid (85 mg Al/kg/day) on gestation days 6 to 15 (Colomina et al. 1994).

Developmental Effects of Aluminum and Maternal Stress

A number of studies in mammals have shown that during pregnancy, maternal stress from restraint, noise, light, and heat among others may be associated with adverse effects on embryo/fetal development (Scialli 1988). Of special concern is the finding that interaction between maternal stress and some chemical teratogens can enhance the developmental toxicity of those chemicals. Since Al is ubiquitous, exposure to this element is in fact unavoidable. This means that pregnant women may be potentially exposed to Al in food, drinking water, soil ingestion, and some medications. They may also be concurrently exposed to various types of stress, either at home or in the workplace. Because both Al and maternal stress during pregnancy can produce adverse developmental effects in mammals, we investigated the developmental toxicity of a combined exposure to Al and maternal stress. The model stressor used was maternal immobilization. Four groups of plug-positive female mice were given intraperitoneal injections of $AlCl_3$ at 37.5 and 75 mg/kg/day on days 6 to 15 of gestation. Two of these groups were also subjected to restraint for 2 h/day during the same gestational days. Maternal toxicity was significantly enhanced by restraint stress at 75 mg $AlCl_3$/kg/day. No signs of embryo/fetal toxicity were observed following exposure to Al, maternal restraint, or combined Al and restraint. However, a significant decrease in fetal body weight, as well as a significant increase in the number of litters with morphologic defects was observed in the group exposed to 75 mg $AlCl_3$/kg/day plus maternal restraint. These results suggest that maternal stress exacerbates Al-induced maternal and developmental toxicity only at high doses of the metal, which are also toxic to the dam (Colomina et al. 1998).

Protective Effects of Chelators on Al-Induced Maternal and Developmental Toxicity

Chelating agents such as desferrioxamine (DFO) and some 3-hydroxypyridin-4-ones can be effective in reducing Al body burdens (Yokel et al. 1996; Gomez et al. 1998a, b). As there were no data on potential chelation therapies to protect pregnant women, infants and children against Al-induced maternal and/or developmental toxicity, the protective activity of DFO and deferiprone (1,2-dimethyl-3-hydroxypyrid-4-one, L1), two efficient chelators for the treatment of Al overload, on Al-induced maternal and developmental toxic effects was evaluated in mice. In a first study, Al chloride was intraperitoneally injected in pregnant mice at 0, 60, 120, and 240 mg/kg/day on gestation days 6 to 15, while DFO was administered subcutaneously at 40 mg/kg/day on days 6 to 18 of gestation. In a previous study (Bosque et al. 1995), we found that the NOAEL for developmental toxicity of parenteral DFO in mice was 176 mg/kg/day. Significant amelioration by DFO of Al-induced maternal toxicity was only noted at 120 mg/kg/day, while no DFO effects were observed on fetal body weight, the only embryo/fetal parameter significantly affected by maternal Al exposure (Albina et al. 1999). The unexpected lack of Al-induced embryotoxicity found in this study could be due to the chelating activity of DFO on Al^{3+} in dam tissues, which would prevent this ion from reaching the embryo. In a second study, a single oral dose of Al nitrate nonahydrate (1327 mg/kg) was given to mice on gestation day 12, the most sensitive time for Al-induced maternal and developmental toxic effects in this species. At 2, 24, 48, and 72 hr thereafter, deferiprone was given by gavage at 0 and 24 mg/kg. Aluminum-induced maternal toxicity was evidenced by significant reductions in body weight gain and food consumption, while developmental toxicity was evidenced by a significant decrease in fetal weight per litter and an increase in the total number of fetuses and litters showing bone retardation. No beneficial effects of deferiprone on these adverse effects could be noted. In contrast, a more pronounced decrease in maternal weight gain, as well as an increase in the number of litters with fetuses showing skeletal variations, were observed in the group given Al and deferiprone (Albina et al. 2000).

As both DFO and deferiprone failed to protect against Al-induced maternal and developmental toxicity, we also investigated whether dietary Si could prevent the toxic effects caused by Al in the pregnant animals. The rationale of this study was based on clear evidence showing that oral silicon can reduce the gastrointestinal absorption of Al and increases its elimination (Yokel et al. 1996; Bellés et al. 1998). The preventive mechanism appears to involve the formation of hydroxyaluminosilicates by the adsorption of silicic acid onto an Al hydroxide template. On gestation days 6 to 15, Al nitrate nonahydrate (398 mg/kg/day) was given by gavage to three groups of pregnant mice, which also received silicon in drinking water at concentrations of 0, 118, and 236 mg/l on days 7 to 18 of gestation. Although silicon administration at 236 mg/l significantly reduced the percentage of Al-induced maternal deaths, abortions and early deliveries, neither 118 nor 236 mg/l of silicon produced significant amelioration on Al-induced fetotoxicity (Bellés et al. 1999).

Conclusions

The results of the studies here reviewed, including those on the enhancement of Al absorption from the gastrointestinal tract by certain dietary constituents, the potential effects of maternal stress on Al-induced maternal and developmental toxic effects, and the lack of an adequate, safe, and effective treatment to protect against these potential adverse effects, indicate the advisablility of avoiding high-dose consumption of Al-containing compounds during gestation and lactation. Data on human studies also suggest this recommendation. The results of Weberg et al. (1986), who performed a small trial to assess whether antacids containing Al were safe during pregnancy, and the findings of Gilbert-Barness et al. (1998), based on data related to the death of a 9-year-old girl who failed to progress developmentally at age 2 months, and whose mother ingested very high amounts of Al hydroxide daily during the entire pregnancy, support the hypothesis that Al exposure during pregnancy can be a developmental hazard.

Uranium

Uranium (U) is a naturally occurring element, the best known use of which in the last 60 years has been as fuel in nuclear power reactors and nuclear weapons. During uranium processing, workers may inhale or ingest some uranium, giving rise to internal contamination, which could result in radiation doses to the body. In addition, if uranium exposure were large enough, chemical toxicity could also occur. Under some circumstances, the chemical toxicity of soluble uranium compounds can even surpass the potential radiotoxic effects. The general population may be exposed to low levels of uranium by inhalation or through the diet, while uranium may be also introduced into drinking water supplies through the mining and milling of uranium ore (Domingo 1995b).

In the early days of the Manhattan Project, a very extensive toxicology program on uranium was carried out (Tannenbaum 1951; Voetglin and Hodge 1953). The principal objectives included the establishment of exposure limits for airborne uranium in the workplace based upon uranium's known chemical renal damage. Although the biokinetics, metabolism, and chemical toxicity of uranium, including the toxic effects of this metal on kidney function, are well established (Wrenn et al. 1985; Domingo 1995b), until recently there was a lack of published observations regarding uranium-induced reproductive and developmental toxic effects (Arfsten et al. 2001; Domingo 2001). In 1987, a program directed at filling the gaps regarding uranium reproductive and developmental toxicity in mammals, as well as concerning possible prevention/amelioration by chelating agents was started in our laboratory. Although uranium can exist in oxidation states +3, +4, +5, or +6, in solution the uranyl ion (UO_2^{2+}) is the most stable species and the form in which this element is present in the mammalian body (Domingo 1995b). Taking this into account, in our studies uranium was administered as uranyl acetate. The results of those studies, together with some additional data reported by other investigators, are summarized here in Table 4.2. The very few data on the effects of depleted uranium (DU) on reproduction that are available from the literature are also discussed. DU is a low-level radioactive waste product of the enrichment of natural uranium with U-235 for reactor fuels or nuclear weapons. In DU, most of the U-235 and U-234 isotopes have been selectively removed through industrial processes, meaning that the radiologic hazard of DU is less than that from natural or enriched uranium. However, DU is also a heavy metal with toxicity being a function of the route of exposure, particle solubility, contact time, and route of elimination. Consequently, although DU exposure can result in both chemical toxicity and toxicity from radioactivity, the chemical toxic effects (mainly on the kidney) occur, in general, at lower exposure levels than the radiologic toxic effects (McDiarmid 2001; Priest 2001).

Reproductive Toxicity of Uranium

Information concerning the reproductive toxicity of uranium is scarce. Most reproductive effects of uranium are based on its chemical nature and properties rather than on its radioactive action. The cytogenetic damage induced by a wide range of concentrations of uranyl fluoride injected into mouse testes was evaluated to determine the frequencies of chromosomal aberrations in spermatogenia and primary spermatocytes. It was observed that the damage depended on the administered dose of uranyl fluoride (Hu and Zhu 1990). Both radiotoxicity and chemical toxicity of uranyl fluoride were considered to be responsible for the adverse effects.

Table 4.2 Embryo/Fetal Toxicity of Uranium: A Summary of Studies

Uranium Compound	Species	Doses (mg/kg/day)	Dosing Period	Route	Toxic Effects	Reference
Uranyl acetate dihydrate	Male mice	10, 20, 40, and 80	64 days before mating	Drinking water	Interstitial alterations at 80 mg/kg/day. Reduction in pregnancy rate at all doses	Llobet et al. (1991)
Uranyl nitrate hexahydrate	Male rats	Not reported	12 months	Diet	Severe degeneration in the testes, depletion of germ cells	Maynard et al. (1953)
Uranyl nitrate hexahydrate	Male rats	0.07	16 weeks	Diet	Decreased testes weight, testicular lesions, necrosis of spermatocytes	Malenchenko et al. (1978)
Uranyl acetate dihydrate	Mice	5, 10, 25, and 50	days 6–15 of gestation	PO	Fetotoxicity including teratogenicity	Domingo et al. (1989b)
Uranyl acetate dihydrate	Mice	0.05, 0.5, 5, and 50	days 13–18 of gestation	PO	Decreases in viability and lactation indices	Domingo et al. (1989c)
Uranyl acetate dihydrate	Mice	5, 10, and 25	Males 60 days and females 14 days before mating	PO	Lower viability indices and reduced growth of the offspring	Paternain et al. (1989)
Uranyl acetate dihydrate	Mice	0.5, 1, and 2	days 6–15 of gestation	SC	Embryo/fetal toxicity including teratogenicity	Bosque et al. (1993a)
Uranyl acetate dihydrate	Mice	4	one of days 9–12 of gestation	SC	Embryo/fetal toxicity; most sensitive effects: day 10	Bosque et al. (1992)
Uranyl acetate dihydrate (+ Tiron)	Mice	4 + Tiron (500, 1000, and 1500)	gestation day 10; Tiron, days 10–13	SC	Developmental toxicity in Tiron-untreated group	Bosque et al. (1993b)

The possibility that chronic uranium exposure of males might affect reproduction in mammals was investigated in our laboratory (Llobet et al. 1991). Male Swiss mice had, as their water source for 64 days, solutions of uranyl acetate dihydrate at concentrations of 0, 0.047, 0.091, 0.170, and 0.420 mg/ml, resulting in doses of 0, 10, 20, 40, and 80 mg/kg/day. To evaluate male fertility, animals were mated with untreated females for 4 days. There was a significant but non-dose–related decrease in the pregnancy rate of these animals (25 to 35% in uranium treated animals vs. 81% in the control group), while body weights were significantly reduced at 80 mg/kg/day (35.8 ± 2.04 g vs. 37.6 ± 2.53 g in the control group). Testicular function/ spermatogenesis was not affected by uranium at any dose, as evidenced by normal testes and epididymis weights and normal spermatogenesis. Histopathologic examination of the testes in mice killed after 64 days of treatment did not reveal any significant difference between controls and uranium-exposed animals in tubule diameter, tubule alterations, and interstitial alterations (focal atrophy, binucleated cells), with the exception of an increase in Leydig cells vacuolization at 80 mg/kg/day. Although these changes might all have contributed to the reduction in pregnancy rate, it is also possible that uranyl acetate treatment for 64 days produced behavioral changes (including a decrease in the libido of those animals), which contributed, in turn, to this reduction. However, in a previous study in which mature male mice were given by gavage 0, 5, 10, or 25 mg/kg/day uranyl acetate dihydrate for 60 days prior to mating with mature virgin female mice, exposed to the same uranium doses for 14 days prior to mating, no adverse effects of uranium on fertility were evident at any dose, while embryolethality was observed at 25 mg/kg/day (Paternain et al. 1989). According to the above results, the NOAEL for reproductive toxicity of uranium is below 10 mg/kg/day, as at that dose the pregnancy rate was significantly diminished (Llobet et al. 1991). The oral LD_{50} of uranyl acetate dihydrate in mice was previously found to be 242 mg/kg, with confidence limits between 155 and 327 mg/kg (Domingo et al. 1987c). Therefore, 5, 10, and 25 mg/kg/day of uranyl acetate dihydrate corresponded approximately to 1/50, 1/25, and 1/ 10 of the acute oral LD_{50} for this compound.

The above findings corroborated earlier studies suggesting the possibility that chronic uranium exposure in males might affect reproduction (Maynard et al. 1953; Malenchenko et al. 1978). While a chronic diet of uranyl nitrate hexahydrate given to rats for 12 months caused severe degeneration in the testes and depletion of germ cells (Maynard et al. 1953), 0.07 mg/kg/ day uranyl nitrate hexahydrate added to the diet of rats for 16 weeks resulted in decreased testes weight, testicular lesions, and necrosis of spermatocytes and spermatogonia (Malenchenko et al. 1978).

With respect to the reproductive effects of DU (depleted uranium), Pellmar and coworkers (1999) assessed in rats the potential health risks associated with chronic exposure to DU. Animals were surgically implanted with DU pellets in muscle at 3 dose levels (low-dose, 4 DU pellets; medium-dose, 10 DU pellets, and high-dose, 20 DU pellets), and uranium distribution was determined over the course of 18 months. These investigators found that although kidney and bone were the primary reservoirs for uranium redistributed from intramuscularly embedded fragments, accumulation in testicles, as well as in brain and lymph nodes, suggested the potential for unanticipated physiologic consequences of uranium exposure through this route.

In recent years, little attention has been paid to the possible effects of uranium exposure on human reproduction. At the time of the original toxicology evaluations during the Second World War, two studies were performed, one of which featured exposure to high levels of the metal, while the other involved only a brief 24-hr exposure (Domingo 2001). Although, in both studies statistically significant effects on reproduction were found (data not shown), the results were not repeated or extended by other investigators. In a study on the sex ratio of offspring of male uranium miners, more female offspring than predicted were noted, suggesting potential alterations in sperm (Muller et al. 1967). More recently, unexpected rates of chromosomal instability and alterations of hormone levels were also found in uranium miners (Zaire et al. 1997).

Embryo/Fetal Toxicity of Uranium

The maternal and developmental toxicity of uranyl acetate dihydrate given by gavage at doses of 0, 5, 10, 25, and 50 mg/kg/day was evaluated in pregnant Swiss mice on gestation days 6 to 15 (Domingo et al. 1989b). Maternal toxicity was evidenced by decreased weight gain and food consumption, as well as elevated relative liver weight at all dose levels tested. Thus, the NOAEL for maternal toxicity was below 5 mg/kg/day. Although there was no evidence of embryolethality at maternally toxic levels, dose-related fetal toxicity, consisting primarily of reduced fetal body weight and body length, and an increased incidence of developmental anomalies, were observed. External malformations and variations included cleft palate and hematomas (dorsal and facial areas). In turn, bipartite sternebrae, reduced ossification of skull and caudal bones, and poor ossification of some hindlimb metatarsals and some proximal forelimb phalanges were the most notable skeletal variations (65 to 100% of affected litters in the uranium-treated groups vs. 22% in the control group). Although various anomalies could be caused by a number of maternal stressors, some of the fetal defects found in this study were also reported to occur independently on maternal toxicity. Consequently, the NOAEL for fetotoxicity including teratogenicity was below 5 mg/kg/day, as some anomalies were observed at this level.

In order to examine whether the developmental toxicity of uranium depends on the route of exposure, the effects of multiple maternal SC injections of uranyl acetate dihydrate (0, 0.5, 1, and 2 mg/kg/day) given from day 6 through 15 of gestation were evaluated in mice (Bosque et al. 1993a). The doses of 0.5, 1, and 2 mg/kg/day were approximately equal to 1/40, 1/20, and 1/10 of the acute SC LD_{50} for uranyl acetate dihydrate. Maternal toxicity occurred in all uranium-treated groups as evidenced primarily by some deaths (0, 1, 2, and 7 deaths at 0, 0.5, 1, and 2 mg/kg/day, respectively) and decreases in body weight gain and body weight at termination. Embryotoxicity was also noted in all uranium-exposed groups. Fetotoxicity was indicated by a significant reduction in fetal weight and significant increases in the incidence of several unossified districts, or by decreased ossification at 1 and 2 mg/kg/day (50 and 100%, respectively, of affected litters vs. 9% in the control group). Cleft palate and bipartite sternebrae were the most notable malformations observed at these doses. According to the conclusions of a review on teratology studies conducted in mice (Khera 1984), the embryotoxic effects detected in our study would be attributable to a direct consequence of uranium-induced maternal toxicity. However, malformations such as cleft palate or some developmental variations were not reported to be defects resulting from maternal toxicity (Khera 1984). Consequently, they probably would be primary effects of the developmental toxicity of uranium. On the basis of these data, the NOAELs for maternal toxicity and for embryotoxicity were below 0.5 mg/kg/day, whereas the NOAEL for teratogenicity was 0.5 mg/kg/day.

The influence of exposure day on the embryo/fetal toxicity of uranium was also examined. Single SC injections of 4 mg/kg of uranyl acetate dihydrate were given to pregnant mice on one of days 9 through 12 of gestation. Dams were killed on day 18 of pregnancy and their uterine contents examined. Although the number of dead and resorbed fetuses, as well as the percentage of postimplantation loss, were significantly increased on any of gestation days 9 through 12, the most sensitive time for the induction of uranium embryotoxicity was gestation day 10. Uranium exposure on days 9 through 12 of gestation also resulted in significant reductions in fetal body weight (0.84 to 0.94 g in the uranium treated groups vs. 1.32 g in the control group) and a high percentage of total skeletal defects. It was concluded that gestation day 10 was the most sensitive time for uranium-induced developmental toxicity in mice (Bosque et al. 1992).

Only a short abstract regarding the maternal and/or developmental effects of DU has been found in the literature. Female rats were exposed at one of five doses (not reported) of DU via surgically implanted pellets and then bred with male rats. On gestation day 20, dams were euthanized and uranium levels in placenta, whole fetus, fetal liver, and maternal kidney were determined. Although an increasing trend of uranium levels in maternal kidney, placenta, and whole fetus with increasing levels of maternal DU implantation was noted, no maternal or fetal toxicity were evident. No adverse effects on maternal weight gain, food consumption, water intake, or histology of the kidney were observed, and parameters such as litter size, pup weight, and sex ratio were not affected by DU exposure (Benson and McBride 1997).

In order to specifically address the association between birth defects, stillbirths, and other adverse outcomes of pregnancy and exposure to uranium from mining and milling operations, a study including 13,329 Navajos born at the Public Health Service/Indian Health Service Hospital (Shiprock, New Mexico) was conducted (Shields et al. 1992). The only statistically significant association between uranium operations and unfavorable birth outcome was identified with the mother living near tailings or mine dumps. The associations between adverse pregnancy outcome and exposure to uranium were weak and attributed to radiation, not to the chemical toxicity of uranium.

Prevention by Chelating Agents of Uranium-Induced Developmental Toxicity

A number of experimental studies have shown that uranium intoxication can be alleviated by administration of chelating agents (Domingo 1995c). Tiron (sodium 4,5-dihydroxybenzene-l,3-disulfonate) and DTPA (diethylenetriaminepentaacetic acid) were found to be the most effective chelators in mobilizing uranium in rats and mice, with DTPA being less effective than Tiron (Domingo et al. 1992).

To determine whether Tiron could ameliorate the developmentally toxic effects of uranium in mice, a series of four Tiron injections was administered IP to pregnant mice immediately after a single SC injection of 4 mg/kg uranyl acetate dihydrate given on day 10 of gestation, and at 24, 48, and 72 hr thereafter (Bosque et al. 1993b). Tiron effectiveness was assessed at 500, 1000, and 1500 mg/kg. In a previous study, the NOAEL for maternal and developmental toxicity of Tiron was found to be 1500 mg/kg/day (Ortega et al. 1991). Dam mortality (20%) was only observed in the group given uranium plus saline (positive control group). In addition, while at the end of the gestation period, maternal body weight in the positive control group was significantly lower than in the uranium-untreated group (negative control), no significant differences between the groups given uranium plus Tiron and the negative control group were seen in maternal weight. It was concluded that treatment with Tiron ameliorated the general condition of the dams. Although uranium-induced embryolethality was not significantly reduced at 500, 1000, or 1500 mg Tiron/kg/day, significant protective effects of Tiron against uranium-induced fetal growth retardation were noted at 1500 mg/kg/day. The protective effects of the drug were probably due to the fact that Tiron reduced the amount of uranium administered to pregnant mice to such low levels, which could partly obviate its developmental toxicity. However, the ability of Tiron to protect the developing mouse fetus against at least some harmful effects of uranium offered only modest encouragement with regard to the possible therapeutic potential for pregnant women exposed to this metal.

Conclusions

The United Nations Scientific Committee on the Effects of Atomic Radiation (UNSCEAR) established that limits for natural uranium in drinking water should be based on its chemical toxicity for the kidney rather than on a hypothetical radiologic toxicity for skeletal tissue. A level of 100 μg u/l of water was chosen as reasonable based on considerations of renal toxicity with the application of a safety factor of 50 to 150 (Wrenn et al. 1985). Consequently, a 70-kg adult consuming 2 l/day of water would not ingest more that 200 μg/day of U. This amount would be equivalent to 0.005 mg/kg/day of uranyl acetate dihydrate. Compared to the NOAEL of uranyl acetate dihydrate for health hazards to the developing pup, 5 mg/kg/day (Domingo et al. 1989c), and the NOAEL for effects of this chemical on reproduction, gestation, and postnatal survival, 5 mg/kg/day (Paternain et al. 1989), a margin of exposure (MOE) of 1000 (MOE: 5 mg/kg/day [NOAEL]/0.005 mg/kg/day [human exposure]) was obtained for the intake of uranium from drinking water. With respect to the fetal toxicity of uranyl acetate dihydrate with a lowest observed adverse effect level (LOAEL) of 5 mg/kg/day (Domingo et al. 1989b), a MOE of 100 (MOE: 0.5 mg/kg/day [LOAEL/10]/0.005 mg/ kg/day [human exposure]) was estimated. Although these MOEs are relatively large, people living near uranium mines or mills may be exposed to higher quantities of the metal and may require individual guidance. Inasmuch as a significant decrease in the pregnancy rate in mice was noted at 10 mg/kg/day uranyl acetate dihydrate (Llobet et al. 1991), investigations are still required to elucidate the mechanism of this effect and whether it may be totally or partially reversible. Studies on the developmental toxicity of DU are also clearly required. Finally, it is important to note that, to date, most studies on uranium-induced developmental toxicity have been performed in mice. Consequently, developmental and reproductive investigations on the effects of uranium in other species of mammals would be also of interest.

Vanadium

Vanadium compounds are extensively used in modern industry expanding from its classic use in steel production to fields such as elements in super-conductive materials. Vanadium is also one of the metallic components contained in crude petroleum oils. Consequently, in addition to a number of industrial activities vanadium is also released into the environment through the combustion of fossil fuels. While occupational exposure to vanadium is not uncommon, in the nonindustrial community vanadium exposure in humans arises primarily from its environmental presence. On the other hand, it has been demonstrated that the inclusion of vanadium compounds in the drinking water of streptozotocin-induced diabetic rats ameliorated some signs of diabetes (Heyliger et al. 1985; Cam et al. 1993). However, because of the serious toxic side effects derived from vanadium treatment (including significant accumulation of vanadium in tissues), oral administration of vanadium compounds should not be considered an adequate alternative therapy to the parenteral use of insulin in diabetic patients. Because of this limitation to the potential use of vanadium in diabetes treatment, toxicologic investigations of this element have been of interest. Bearing on this question, hematologic and biochemical alterations, renal toxicity, immunotoxicology, and mutagenicity were observed following exposure to vanadium compounds (Domingo et al. 1995d; Domingo 2000, 2002). In recent years, it has also been shown that vanadium may cause reproductive and developmental toxicity in rats, mice, and hamsters (Leonard and Gerber 1994; Domingo 1996, 2002).

Vanadate (VO_3^-, V^{+5}) and vanadyl (VO^{+2}, V^{+4}) are the vanadium oxidation states of main biological interest. It is well established that both vanadate and vanadyl may cause a number of adverse toxic effects in mammals depending on circulating levels. Among these, reproductive and developmental toxicity have been demonstrated to occur following vanadium exposure. The effects of vanadium on reproduction and development in mammals are here reviewed. As for aluminum and uranium, special attention is paid to the studies performed in our laboratory.

Reproductive Toxicity of Vanadium

Information concerning the reproductive toxicity of vanadium is rather scarce. Kamboj and Kar (1964) investigated in rats and mice the antitesticular effects of a number of water-soluble salts of metals and rare earths, including vanadium sulfate (hypovanadous). A single intratesticular injection of 0.08 mmol/kg in rats reduced mean testis weight and produced total necrosis of the testes within 2 days after vanadium exposure. Vanadium sulfate caused severe damage to the seminiferous epithelium with exfoliation and lysis of the cellular elements. However, there was no effect on the residual spermatozoa in the ductus deferens. Although subcutaneous injection of vanadium sulfate (0.08 mmol/kg) in mice also significantly reduced mean testis weight, no necrotic changes were seen in the testis of those animals. Recently, Morgan and El-Tawil (2003) administered to male and female rats 200 ppm ammonium metavanadate in drinking water for 70 and 14 days before mating, respectively. The fertility was significantly reduced in both groups, with more pronounced suppressive effects in the male treated group.

In a previous investigation, the gonadotoxic effects of vanadium were examined in rats (the route of exposure and the chemical form of vanadium were not specified) (Roschin et al. 1980). These effects were characterized by disturbances of spermatogenesis, of the function of spermatozoa (reduced motility and increased number of dead spermatozoa), and by morphologic disturbances in the spermatozoa and the seminiferous epithelium. Vanadium also accumulated in the testes, and fertilization of female rats did not occur after copulation with males exposed to vanadium at a dose of $1/20$ LD_{50} for a period of 20 days.

In a study performed in our laboratory (Domingo et al. 1986), adult male rats were orally exposed to sodium metavanadate (V+5) at doses of 0, 5, 10, and 20 mg/kg/day for 60 days before mating with females that received the same vanadate doses during the 14 days just prior to mating. Treatment of female rats with vanadate was continued during the periods of gestation and lactation. Although no significant adverse effects on fertility, reproduction, and parturition were observed, the development of the offspring (body weight, body length, and tail length) was significantly decreased from birth and during the entire lactation period at 5, 10, and 20 mg/kg/day.

Sodium metavanadate was also given to male mice for 64 days before mating with vanadate-untreated females. Sodium metavanadate was administered in the drinking water at concentrations of 0, 0.10, 0.20, 0.30, and 0.40 mg/ml to obtain dosages of 0, 20, 40, 60, and 80 mg/kg/day. A significant decrease in the pregnancy rate was seen at 60 mg/kg/day (43.8% vs. 81.3% in the control group) and 80 mg/kg/day (62.5% vs. 81.3%). Decreased body and epididymis weight were observed in the 80 mg/kg/day group, but testicular weights were not altered by vanadium treatment at any dose administered. It was previously shown that vanadium crosses the blood-testis barrier and accumulates in the testes (Parker et al. 1980). However, although in our study sperm count was significantly diminished at 40, 60, and 80 mg/kg/day, sperm motility was unaffected. In turn, histopathologic examination revealed that the testes were normal and the epididymis contained normal-appearing sperm. The NOAEL was 40 mg/kg/day (7.54 mg V/kg/day) (Llobet et al. 1993).

Altamirano and associates (1991) evaluated the effects of vanadium pentoxide (V+5) administered to prepubertal rats. Vanadium pentoxide (12.5 mg/kg) injected IP in male and female rats every 2 days from birth to day 21, or to female rats from day 21 to the day of first vaginal oestrus, did not cause significant differences in the age of vaginal opening, or in the weight of ovaries, uterus, addrenals, and pituitary. Notwithstanding, the ovulation rate was lower in vanadium treated animals (8 out of 15) than in controls (18 out of 19), but no differences were observed in the ovulation rate of newborn rats or for 21 days after birth. The number of ova shed by ovulating animals was similar in control and treated animals. Such results could indicate that vanadium pentoxide affected only ovulation, but did not modify the secretion of oestrogen. In male rats, an increase in the weight of seminal vesicles, thymus, and submandibular glands could be observed. It was concluded that as previously reported for other metals, vanadium toxicity in prepubertal rats is higher in males than in females (Altamirano, et al., 1991).

The above studies show that there are remarkable differences in the reproductive toxicity of vanadium, which depend on the route of exposure. The damage to the testis after intratesticular injection could be due to the formation of vanadium complexes with a number of ligands containing sulfur, nitrogen, or oxygen, which are integral parts of molecules of biologic significance, including enzyme systems, as well as action on the cell membrane causing its dissolution. By contrast, oral exposure to vanadate (V+5) does not cause adverse effects on fertility or testicular function at the levels usually ingested by humans through diet and drinking water. Taking as a basis a safe intake of 200 μg V/person/day (about 10 times the average vanadium via diet), a person of 60 kg weight would ingest 3.33 μg V/kg/day, giving a safety factor of more than 2000 (7540 μg V kg/day; 3.33 μg V/kg/day).

Embryo/Fetal Toxicity of Vanadium

Vanadium is known to cross the placenta and accumulate in the fetus, especially in the fetal skeleton (Domingo 1996). A mild transplacental effect of
vanadium on development was observed in the hamster when pregnant
animals received ammonium vanadate (0.47, 1.88, and 3.75 mg/kg/day) by
IP injection on days 5 through 10 of gestation. Although no significant
maternal toxicity was apparent at any of the dose levels, skeletal anomalies
included micrognathia, supernumerary ribs, and alterations in sternebral ossification. In spite of these anomalies, it was concluded that the low incidence
and lack of dose-dependent response would not allow a definite assignation
of teratogenicity to ammonium vanadate exposure (Carlton et al. 1982).

Edel and Sabbioni (1989) injected IV low doses of [48]V-labeled pentavan-
adate (V+5: 0.1 µg V/animal) to pregnant rats on gestation day 12. Significant
vanadium concentrations were found in the liver, intestine, and kidneys of
the fetuses, showing that vanadate passes the placental barrier and is metabolized in the fetus. Vanadyl (V+4) is also capable of crossing the placenta
and reaching the fetus. When vanadyl sulfate pentahydrate was administered by gavage to pregnant mice at doses of 0, 37.5, 75, and 150 mg/kg/
day on days 6 to 15 of pregnancy, there was a significant increase in vanadium concentrations in the placenta (Paternain et al. 1990). In that study, as
well as in the report by Edel and Sabbioni (1989), the amounts of vanadium
found in the placental unit were remarkably higher than those found in
fetuses. This suggests that the placenta represents a partial barrier to vanadium exposure. The transplacental transport of vanadium might occur in
different ways: in the form of strongly complexed transferrin, as a weak
bicomplex with albumin, or as a more simple compound.

In an earlier study, Roshchin and associates (1980) observed that administration of vanadium (oxidation state was not reported) to rats on gestational
days 21 and 22 caused an accumulation of the metal in the placenta, but did
not penetrate the placental barrier to any significant degree. However, Roshchin
et al. (1980) found some embryotoxic effects of vanadium in rats characterized
by a considerable increase in mortality of embryos due to an enhancement in
preimplantation death of embryos following vanadium administration on day
10 of pregnancy. The number of fetuses from vanadium-treated dams was
approximately one-half of that observed in control animals. In turn, the IV
administration of 0.15 ml of vanadium pentoxide (V+5) 1 mM to mice on day
8 of pregnancy reduced skeletal ossification in 71% of the fetuses without
increasing the number of nonviable implants (Wide 1984). A broken spinal cord
was also seen in 9% of the fetuses examined on day 17 of gestation.

As for most toxic elements, the importance of the route of exposure on
the embryo/fetal toxicity of vanadium has been clearly shown. Thus, while
sodium metavanadate was neither embryolethal nor teratogenic in rats at
doses of 20 mg/kg/day or lower when it was orally administered on gestation days 6 to 14 (Paternain et al. 1987), an increased number of resorptions
and dead fetuses, an increased percentage of postimplantation loss, as well
as an increase in the incidence of cleft palate was found when metavanadate

was given IP to mice on days 6 to 15 of pregnancy at 4 or 8 mg/kg/day (Gomez et al. 1992). Although in the study by Gomez et al. (1992) maternal toxicity was also noted at 1, 4, and 8 mg/kg/day, as evidenced by decreased weight gain during vanadate treatment, the pattern of embryo/fetal toxicity would be the result of the direct contact of vanadium with the embryonic or fetal tissues rather than an effect derived from maternal toxicity.

The chemical form of the vanadium compound is another key factor in explaining the differences found in the results of various developmental toxicity studies of vanadium. Sodium orthovanadate (V+5) given by gavage to mice on gestational days 6 to 15 at doses of 0, 7.5, 15, and 30 mg/kg/d caused maternal toxicity at 15 and 30 mg/kg/day evidenced by some deaths and/or reduced body weight gain and food consumption (Sanchez et al. 1991). Embryolethality and teratogenicity were not observed at maternally toxic doses and below, but fetal toxicity (significant delay in the ossification of some skeletal districts) was seen at 30 mg/kg/day. It was concluded that orthovanadate treatment would result in slight fetal growth retardation in mice only in the presence of well-defined maternal toxicity. In turn, oral administration of vanadyl (V+4) sulfate pentahydrate (0, 37.5, 75, and 150 mg/kg/day) to pregnant mice on days 6 to 15 of gestation resulted in maternal toxicity, embryotoxicity, and fetotoxicity (including teratogenicity) at all dose levels tested (Paternain et al. 1990). Cleft palate and micrognathia were the major gross malformations seen at 150 mg/kg/day.

On the other hand, IP injection of vanadium pentoxide (V+5) (8.5 mg/kg/day) in mice on days 6 to 15 of gestation did not cause significant adverse effects in the number of live and dead fetuses, fetal implants, or resorptions. However, the weight of the fetuses and the number of ossification centers in forelimbs and hindlimbs were significantly lower in the vanadium-treated animals. Limb shortening was the most frequent alteration, while no maternal toxicity was observed (Altamirano et al. 1993). Because vanadium is a potent inhibitor of DNA and protein synthesis, and affects several metabolic processes, these effects might be responsible for the vanadium-induced developmental toxicity.

When the influence of exposure day on the embryo/fetal toxicity of vanadium was evaluated in pregnant mice, it was found that the conceptus can be adversely affected by IP exposure to sodium metavanadate (25 mg/kg) on any of gestational days 9 to 12. Notwithstanding, the most sensitive time for vanadate-induced embryotoxicity was day 12 of pregnancy, whereas the highest percentage of total skeletal defects (82.3%) was also found on gestational day 12 followed by day 9 (Bosque et al. 1993c).

The differences observed in vanadium-induced developmental toxicity studies may be explained considering the different routes of vanadium exposure, the different dose levels of vanadium administered, the different dosing periods/days of exposure, the notable chemical differences among the vanadium compounds tested, or the different oxidation states of vanadium. The results of various studies on the embryo/fetal toxicity of vanadium are summarized in Table 4.3.

Table 4.3 Reproductive Toxicity of Vanadium: A summary of Studies

Vanadium Compound	Species	Doses	Dosing Period	Route	Toxic Effects	Reference
Vanadyl sulfate	Male rats	0.08 mmol/kg	Single dose	Intratesticular	Reduction in testes weight, total necrosis of testes within 2 days after exposure	Kamboj and Kar (1964)
Vanadyl sulfate	Male mice	0.08 mmol/kg	Single dose	SC	Reduction in testes weight	Kamboj and Kar (1964)
Sodium metavanadate	Rats	5, 10, and 20 mg/kg/day	Males, 60 days and females 14 days before mating	PO	No adverse effects on fertility and reproduction	Domingo et al. (1986)
Vanadium pentoxide	Rats	12.5 mg/kg	Every 2 days; males from birth to day 21 and females from day 21 to first vaginal estrus	IP	Significant differences in the age of vaginal opening, and the weight of ovaries, uterus, addrenals, and pituitary. Males: increase in seminal vesicle, thymus, and submandibular gland weight	Altamirano et al. (1991)
Sodium metavanadate	Male mice	20, 40, 60, and 80 mg/kg/day	64 days before mating	Drinking water	Reduced body and epididymis weight at 80 mg/kg/day; decrease in sperm count at 40, 60, and 80 mg/kg/day	Llobet et al. (1993)
Sodium metavanadate	Rats	0.25 mg/ml	Gestation period	Drinking water	Reduction in the number of live fetuses	Ganguli et al. (1994b)
Ammonium metavanadate	Rats	200 ppm	Males 70, and females 14 premating days, and during mating	Drinking water	Adverse effects on fertility and reproduction	Morgan and El-Tawil (2003)

Prevention of Vanadium-Induced Embryo/Fetal Toxicity

The antidotal efficacy of a number of chelating agents on acute and chronic vanadium toxicity has been evaluated in rats and mice. While chelators such as 2,3-dimercaptopropanol (BAL) or sodium ethylenediaminetetraacetate (EDTA) were reported to be largely ineffective, sodium 4,5-dihydroxyben-zene-1,3- disulfonate (Tiron), ascorbic acid, and desferrioxamine, were effective agents for potential use in the treatment of acute and chronic vanadium intoxication (Domingo 1995c). When Tiron was administered SC (250, 500, and 1000 mg/kg) at 0, 24, 48, and 72 hr following IP injection of 25 mg/kg sodium metavanadate on day 12 of gestation, a significant decrease in the number of resorbed fetuses, an increase in the mean fetal weight, and a reduction in the incidence of skeletal variations caused by metavanadate could be observed (Domingo et al. 1993). According to these results, Tiron would offer encouragement with regard to its therapeutic potential for pregnant women exposed to vanadate.

Vanadium Effects on Reproduction and Fetal Development in Diabetes

Manci et al. (1989) found significantly lower placental levels of vanadium in cases of gestational diabetes mellitus (7.62 µg/g) than in matched controls (8.73 µg/g), which was attributed to decreased intake or absorption, and increased utilization or excretion of vanadium. They speculated that there might be increased binding of vanadium to maternal tissues in human diabetes mellitus when insulin is deficient. In relation to this hypothesis, *in vitro* studies showed that vanadate enhances insulin-receptor binding in placenta from women with gestational diabetes mellitus (Al-Attas et al. 1995).

Because of the interest in the potential use of vanadium in diabetes treatment during pregnancy, the effect of oral vanadate on overall reproductive efficiency was evaluated in diabetic and nondiabetic rats (Ganguli et al. 1994a). Vanadate was administered in the drinking water of those animals at concentrations of 0.25 and 0.50 mg/ml. While vanadate was ineffective in normalizing blood glucose in pregnant diabetic rats, it was also found that vanadium treatment had a significant negative impact on the rate of conception and the ability to sustain pregnancy to term in both diabetic and nondiabetic rats. The reduction in the rate of conception might have been due to a change in the estrous cycle; gross examination of the reproductively incompetent dams revealed the frequent presence of polycystic ovaries. Postmortem examination of the uteri from impregnated dams that failed to carry pregnancy to term showed in each case a highly vascularized and fluid-filled uterus, indicating that although these animals conceived, they had an early spontaneous abortion.

In another study, healthy and diabetic pregnant rats received 0.25 mg/ml of sodium metavanadate in drinking water during gestation. The presence of vanadium in maternal blood had a negative effect on fetal development, markedly reducing the number of live fetuses per dam in both healthy and diabetic pregnant animals (Ganguli et al. 1994b). This toxicity could be a consequence of transplacental transfer of a significant amount of vanadate from the maternal to the fetal compartments. In turn, the negative effect of blood vanadium on fetal growth might be a consequence of the generation of excess free radical and reactive species, because it has been documented that high levels of these species can result in embryo death.

According to those results, oral vanadate treatment is toxic and ineffective during diabetic pregnancies, reducing reproductive capacity and interfering with fetal growth and development in both healthy and diabetic animals. Consequently, this treatment would be contraindicated in the management of diabetes in females of childbearing age.

Conclusions

Vanadate (V+5) and vanadyl (V+4), the vanadium oxidation states of biologic interest may cause a number of adverse effects in mammals, depending on circulating vanadium levels. With regard to the effects of vanadium on reproduction and gestation, it is well established that the degree of vanadium toxicity depends on a series of factors such as the chemical form of the specific vanadium compound, the oxidation state of vanadium, the route of exposure, and the period of dosing, as well as the dose of vanadium administered. While there is a lack of information on the reproductive and developmental toxicity of vanadium following inhalation exposure, vanadium has been shown to be a reproductive and embryo/fetal toxicant when given orally. However, toxic effects of vanadate and vanadyl were observed only at dose levels remarkably higher than the amounts of vanadium usually ingested through the diet. Consequently, it seems evident that vanadium would not pose a risk for adverse effects in people under common environmental and nutritional conditions of exposure. Because vanadate impairs reproductive capacity and the ability to sustain pregnancy to term in pregnant diabetic rats, this regimen would be contraindicated in the management of pregnant diabetic females.

References

Al-Attas, O.S., Al-Dagheri, N.M., and Vigo, N.T. (1995) Vanadate enhances insulin-receptor binding in gestational diabetic human placenta. *Cell Biochem Funct*, 13, 9–14.

Albina, M.L., Bellés, M., Sánchez, D.J., and Domingo, J.L. (1999) Prevention by desferrioxamine of aluminum-induced maternal and developmental toxic effects in mice. *Trace Elem Electr*, 16, 192–198.

Albina, M.L., Bellés, M., Sánchez, D.J., and Domingo, J.L. (2000) Evaluation of the protective activity of deferiprone, an aluminum chelator, on aluminum-induced developmental toxicity in mice. *Teratology*, 62, 86–92.

Alfrey, A.C. (1985) Gastrointestinal absorption of aluminum. *Clin Nephrol*, 24, S84–S87.

Alfrey, A.C. (1986) Dialysis encephalopathy. *Kidney Int*, 29, 853–857.

Altamirano, M., Alvarez, L., and Roldan, E. (1993) Cytogenetic and teratogenic effects of vanadium pentoxide on mice. *Med Sci Res*, 21, 711–713.

Altamirano M., Ayala M.E., Flores A., Morales L., Dominguez R. (1991) Sex differences in the effects of vanadium pentoxide administration to prepubertal rats. *Med Sci Res*, 19, 825–826.

Arfsten, D.P., Still, K.R., and Ritchie, G.D. (2001) A review of the effects of uranium and depleted uranium exposure on reproduction and fetal development. *Toxicol Ind Health*, 17, 180–191.

Bataineh, H., Al-Hamood, M.H., and Elbetieha, A.M. (1998) Assessment of aggression, sexual behavior and fertility in adult male rats following long-term ingestion of four industrial metal salts. *Hum Exp Toxicol*, 17, 570–576.

Bellés, M., Albina, M.L., Sanchez, D.J., Corbella, J., and Domingo, J.L. (2001) Effects of oral aluminum on essential trace element metabolism during pregnancy. *Biol Trace Elem Res*, 79, 67–81.

Bellés, M., Albina, M.L., Sánchez, D.J., and Domingo, J.L. (1999) Lack of protective effects of dietary silicon on aluminum-induced maternal and developmental toxicity in mice. *Pharmacol Toxicol*, 85, 1–6.

Bellés, M., Sánchez, D.J., Gómez, M., Corbella, J., and Domingo, J.L. (1998) Silicon reduces aluminum accumulation in rats: Relevance to the aluminum hypothesis of Alzheimer's disease? *Alzheimer Dis Assoc Disord*, 12, 83–87.

Benett, R.W., Persaud, T.V.N., and Moore, K.L. (1975) Experimental studies on the effects of aluminum on pregnancy and fetal development. *Anat Anz Bd*, 138, 365–378.

Benson, K.A. and McBride, S.A. (1997) Uranium levels in the fetus and placenta of female rats implanted with depleted uranium pellets prior to breeding. *Toxicologist*, 36, 258.

Bosque, M.A., Domingo, J.L., and Corbella, J. (1992) Embryofetotoxicity of uranium in mice: Variability with the day of exposure. *Rev Toxicol*, 9, 107–110 (in Spanish).

Bosque, M.A., Domingo, J.L., and Corbella, J. (1995) Assessment of the developmental toxicity of deferoxamine in mice. *Arch Toxicol*, 69, 467–471.

Bosque, M.A., Domingo, J.L., Llobet, J.M., and Corbella, J. (1993a) Embryotoxicity and teratogenicity of uranium in mice following subcutaneous administration of uranyl acetate. *Biol Trace Elem Res*, 36, 109–118.

Bosque, M.A., Domingo, J.L., Llobet, J.M., and Corbella, J. (1993b) Effectiveness of sodium 4,5- dihydroxybenzene-1,3,disulfonate (Tiron) in protecting against uranium-induced developmental toxicity in mice. *Toxicology*, 79, 149–156.

Bosque, M.A., Domingo, J.L., Llobet, J.M., and Corbella, J. (1993c) Variability in the embryotoxicity and fetotoxicity of vanadate with the day of exposure. *Vet Hum Toxicol*, 35, 1–3.

Cam, N.C., Pederson, R.A., Brownsey, R.W., and McNeill, J.H. (1993) Long-term effectiveness of oral vanadyl sulfate in streptozotocin-diabetic rats. *Diabetologia*, 36, 218–224.

Carlton, B.D., Beneke, M.B., and Fisher, G.L. (1982) Assessment of teratogenicity of ammonium vanadate using Syrian golden hamsters. *Environ Res*, 29, 256–262.

Colomina, M.T., Esparza, J.L., Corbella, J., and Domingo, J.L. (1998) The effect of maternal restraint on developmental toxicity of aluminum in mice. *Neurotoxicol Teratol*, 20, 651–656.

Colomina, M.T., Gómez, M., Domingo, J.L., Llobet, J.M., and Corbella, J. (1992) Concurrent ingestion of lactate and aluminum can result in developmental toxicity in mice. *Res Commun Chem Pathol Pharmacol*, 77, 95–106.

Colomina, M.T., Gómez, M., Domingo, J.L., and Corbella, J. (1994) Lack of maternal and developmental toxicity in mice given high doses of aluminum hydroxide and ascorbic acid during gestation. *Pharmacol Toxicol*, 74, 236–239.

Cranmer, J.M., Wilkins, J.D., Cannon, D.J., and Smith, L. (1986) Fetal-placental-maternal uptake of aluminum in mice following gestational exposure: Effect of dose and route of administration. *Neurotoxicology*, 7, 601–608.

Domingo, J.L., Paternain, J.L., Llobet, J.M., and Corbella, J. (1986) Effects of vanadium on reproduction, gestation, parturition and lactation in rats upon oral administration. *Life Sci*, 39, 819–824.

Domingo, J.L., Paternain, J.L., Llobet, J.M., and Corbella, J. (1987a) The effects of aluminum ingestion on reproduction and postnatal survival in rats. *Life Sci*, 41, 1127–1131.

Domingo, J.L., Paternain, J.L., Llobet, J.M., and Corbella, J. (1987b) Effects of oral aluminum administration on prenatal and postnatal progeny development in rats. *Res Commun Chem Pathol Pharmacol*, 57, 129–132.

Domingo, J.L., Llobet, J.M., Tomas, J.M., and Corbella, J. (1987c) Acute toxicity of uranium in rats and mice. *Bull Environ Contam Toxicol*, 39, 168–174.

Domingo, J.L., Gómez, M., Bosque, M.A., and Corbella, J. (1989a) Lack of teratogenicity of aluminum hydroxide in mice. *Life Sci*, 45, 243–247.

Domingo, J.L., Paternain, J.L., Llobet, J.M., and Corbella, J. (1989b) The developmental toxicity of uranium in mice. *Toxicology*, 55, 143–152.

Domingo, J.L., Ortega, A., Paternain, J.L., and Corbella, J. (1989c) Evaluation of the perinatal and postnatal effects of uranium in mice upon oral administration. *Arch Environ Health*, 44, 395–398.

Domingo, J.L., Gómez, M., Llobet, J.M., and Corbella, J. (1991a) Influence of some dietary constituents on aluminum absorption and retention in rats. *Kidney Int*, 39, 598–601.

Domingo, J.L., Gómez, M., Llobet, J.M., and Richart, C. (1991b) Effect of ascorbic acid on gastrointestinal aluminum absorption. *Lancet*, 338, 1467.

Domingo, J.L., Colomina, M.T., Llobet, J.M., Jones, M.M., Singh, P.K., and Campbell, R.A. (1992) The action of chelating agents in experimental uranium intoxication in mice; variations with structure and time of administration. *Fundam Appl Toxicol*, 19, 350–357.

Domingo, J.L., Luna, M., Bosque, M.A., and Corbella, J. (1993) Prevention by Tiron (sodium 4,5-dihydroxybenzene-1,3-disulfonate) of vanadate-induced developmental toxicity in mice. *Teratology*, 48, 133–138.

Domingo, J.L. (1994) Metal-induced developmental toxicity in mammals: A review. *J Toxicol Environ Health*, 42, 123–141.

Domingo, J.L., Gómez, M., Llobet, J.M., del Castillo, D., and Corbella, J. (1994) Influence of citric ascorbic and lactic acids on the gastrointestinal absorption of aluminum in uremic rats. *Nephron*, 66, 108–109.

Domingo, J.L. (1995a) Reproductive and developmental toxicity of aluminum: A review. *Neurotoxicol Teratol*, 17, 515–521.

Domingo, J.L. (1995b) Chemical toxicity of uranium. *Toxicol Ecotoxicol News*, 2, 74–78.

Domingo, J.L. (1995c) Prevention by chelating agents of metal-induced developmental toxicity. *Reprod Toxicol*, 9, 105–113.

Domingo, J.L., Gomez, M., Sanchez, D.J., Llobet, J.M., and Keen, C.L. (1995) Toxicology of vanadium compounds in diabetic rats: The action of chelating agents on vanadium accumulation. *Mol Cell Biochem*, 153, 233–240.

Domingo, J.L. (1996) Vanadium: A review of the reproductive and developmental toxicity. *Reprod Toxicol*, 10, 175–182.

Domingo, J.L. (1997) Metal-induced developmental toxicity in mammals, in *Advances in Environmental Control Technology*, Cheremisinoff, P., Ed., Gulf Publishing Co., Houston, 395–414.

Domingo, J.L. (2000) Vanadium and diabetes. What about vanadium toxicity? *Mol Cell Biochem*, 203, 185–187.

Domingo, J.L. (2001) Reproductive and developmental toxicity of natural and depleted uranium: A review. *Reprod Toxicol*, 15, 603–609.

Domingo, J.L. (2002) Vanadium and tungsten derivatives as antidiabetic agents: A review of their toxic effects. *Biol Trace Elem Res*, 88, 97–112.

Domingo, J.L. (2003) Aluminum toxicology, in *Encyclopaedia of Food Sciences and Nutrition*, 2nd ed., Caballero, B., Trugo, L., and Finglas, P.M., eds., Academic Press, London, 160–166.

Edel, J. and Sabbioni, E. (1989) Vanadium transport across placenta and milk of rats to the fetus and newborn. *Biol Trace Elem Res*, 22, 265–275.

Ernhart, C.B. (1992) A critical review of low-level prenatal lead exposure in the human: 1. Effects on the fetus and newborn. *Reprod Toxicol*, 6, 9–19.

Ganguli, S., Reuland, D.J., Franklin, L.A., and Tucker, M. (1994a) Effect of vanadate on reproductive efficiency in normal and streptozotocin-treated diabetic rats. *Metabolism*, 43, 1384–1388.

Ganguli, S., Reuland, D.J., Franklin, L.A., Deakins, D.D., Johnston, W.J., and Pasha, A. (1994b) Effects of maternal vanadate treatment on fetal development. *Life Sci*, 55, 1267–1276.

Gilbert-Barness, E., Barness, L., Wolff, J., and Harding, C. (1998) Aluminum toxicity. *Arch Pediatr Adolesc Med*, 152, 511–512.

Golding, J., Rowland, A., Greenwood, R., and Lunt, P. (1991) Aluminum sulfate in water in north Cornwall and outcome of pregnancy. *Br J Med*, 302, 1175–1177.

Golub, M.S. (1994) Maternal toxicity and the identification of inorganic arsenic as a developmental toxicant. *Reprod Toxicol*, 8, 283–295.

Golub, M.S. and Domingo, J.L. (1996) What we know and what we need to know about developmental aluminum toxicity. *J Toxicol Environ Health*, 48, 585–597.

Golub, M.S. and Domingo, J.L. (1998) Fetal aluminum accumulation. *Teratology*, 58, 225–226.

Golub, M.S., Macintosh, M.S., and Baumrind, N. (1998) Developmental and reproductive toxicity of inorganic arsenic: Animal studies and human concerns. *J Toxicol Environ Health B*, 1, 199–241.

Gómez, M., Bosque, M.A., Domingo, J.L., Llobet, J.M., and Corbella, J. (1990) Evaluation of the maternal and developmental toxicity of aluminum from high doses of aluminum hydroxide in rats. *Vet Hum Toxicol*, 32, 545–548.

Gómez, M., Domingo, J.L., and Llobet, J.M. (1991) Developmental toxicity evaluation of oral aluminum in rats: Influence of citrate. *Neurotoxicol Teratol*, 13, 323–328.

Gómez, M., Esparza, J.L., Domingo, J.L., Corbella, J., Singh, P.K., and Jones, M.M. (1998a) Aluminum mobilization: A comparative study of a number of chelating agents in rats. *Pharmacol Toxicol*, 82, 295–300.

Gómez, M., Esparza, J.L., Domingo, J.L., Singh, P.K., and Jones, M.M. (1998b) Comparative aluminum mobilizing actions of deferoxamine and four 3-hydroxy-pyridin-4-ones in aluminum-loaded rats. *Toxicology*, 130, 175–181.

Gómez, M., Sanchez, D.J., Domingo, J.L., and Corbella, J. (1992) Embryotoxic and teratogenic effects of intraperitoneally administered metavanadate in mice. *J Toxicol Environ Health*, 37, 47–56.

Guo, C., Huang, C., Chen, S., and Wang-Hsu, G. (2001) Serum and testicular testosterone and nitric oxid products in aluminum-treated rats. *Environ Toxicol Pharmacol*, 10, 53–60.

Heyliger, C.E., Tahiliani, A.G., and McNeill, J.H. (1985) Effect of vanadate on elevated blood glucose and depressed cardiac performance of diabetic rats. *Science*, 227, 1474–1477.

Hu, Q. and Zhu, S. (1990) Induction of chromosomal aberrations in male mouse germ cells by uranyl fluoride containing enriched uranium. *Mutat Res*, 244, 209–214.

Kamboj, V.P. and Kar, A.B. (1964) Antitesticular effect of metallic and rare earth salts. *J Reprod Fertil*, 7, 21–28.

Khera, K.S. (1984) Maternal toxicity. A possible factor in fetal malformations in mice. *Teratology*, 29, 411–416.

Krasovskii, G.N., Vasukovich, L.Y., and Chariev, O.G. (1979) Experimental study of biological effects of lead and aluminum following oral administration. *Environ Health Perspect*, 30, 47–51.

Leonard, A. and Gerberm G.B. (1994) Mutagenicity, carcinogenicity and teratogenicity of vanadium compounds. *Mutat Res*, 317, 81–88.

Llobet, J.M., Colomina, M.T., Sirvent, J.J., Domingo, J.L., and Corbella, J. (1993) Reproductive toxicity evaluation of vanadium in male mice. *Toxicology*, 80, 199–206.

Llobet, J.M., Colomina, M.T., Sirvent, J.J., Domingo, J.L., and Corbella, J. (1995) Reproductive toxicology of aluminum in male mice. *Fundam Appl Toxicol*, 25, 45–51.

Llobet, J.M., Domingo, J.L., Gómez, M., Tomas, J.M., and Corbella, J. (1987) Acute toxicity studies of aluminum compounds: Antidotal efficacy of several chelating agents. *Pharmacol Toxicol*, 60, 280–283.

Llobet, J.M., Sirvent, J.J., Ortega, A., and Domingo, J.L. (1991) Influence of chronic exposure to uranium on male reproduction in mice. *Fundam Appl Toxicol*, 16, 821–829.

Malenchenko, A.F., Barkun, N.A., and Guseva, G.F. (1978) Effect of uranium on the induction and course of experimental autoimmune orchitis and thyroiditis. *J Hyg Epidemiol Microbiol Immunol*, 22, 268–277.

Manci, E.A., Coffin, C.M., Smith, S.M., and Ganong, C.A. (1989) Placental vanadium in gestational diabetes mellitus. *Placenta*, 10, 417–425.

Maynard, E.A., Downs, W.L., and Hodge, H.C. (1953) Oral toxicity of uranium compounds, in *Pharmacology and Toxicology of Uranium*, Voegtlin, C., Hodge, H.C., eds., McGraw-Hill, New York, 1221–1369.

McDiarmid, M.A. (2001) Depleted uranium and public health. *Br Med J*, 322, 123–124.

Morgan, A.M. and El-Tawil, O.S. (2003) Effects of ammonium metavanadate on fertility and reproductive performance of adult male and female rats. *Pharmacol Res*, 47, 75–85.

Muller, C., Ruzicka, L., and Bakstein, J. (1967) The sex ratio in the offspring of uranium miners. *Acta Univ Carol* (Praha), 13, 599–603.

Ortega, A., Sanchez, D.J., Domingo, J.L., Llobet, J.M., and Corbella, J. (1991) Developmental toxicity evaluation of Tiron (sodium 4,5-dihydroxybenzene-1,3,disulfonate) in mice. *Res Commun Chem Pathol Pharmacol*, 73, 97–106.

Parker, R.D.R., Sharma, R.P., and Oberg, S.G. (1980) Distribution and accumulation of vanadium in mice tissues. *Arch Environ Contam Toxicol*, 9, 393–403.

Partridge, N.A., Regnier, F.E., White, J.L., and Hem, S.L. (1989) Influence of dietary constituents on intestinal absorption of aluminum. *Kidney Int*, 35, 1413–1417.

Paternain, J.L., Domingo, J.L., Gomez, M., Ortega, A., and Corbella, J. (1990) Developmental toxicity of vanadium in mice after oral administration. *J Appl Toxicol*, 10, 181–186.

Paternain, J.L., Domingo, J.L., Llobet, J.M., and Corbella, J. (1987) Embryotoxic effects of sodium metavanadate administered to rats during organogenesis. *Rev Esp Fisiol*, 43, 223–228.

Paternain, J.L., Domingo, J.L., Llobet, J.M., and Corbella, J. (1988) Embryotoxic and teratogenic effects of aluminum nitrate on rats upon oral administration. *Teratology*, 38, 253–257.

Paternain, J.L., Domingo, J.L., Ortega, A., and Llobet, J.M. (1989) The effects of uranium on reproduction, gestation, and postnatal survival in mice. *Ecotoxicol Environ Safety*, 17, 291–296.

Pellmar, T.C., Fuciarelli, A.F., Ejnik, J.W., Hamilton, M., Hogan, J., Strocko, S., Emond, C., Mottaz, H.M., and Landauer, M.R. (1999) Distribution of uranium in rats implanted with depleted uranium pellets. *Toxicol Sci*, 49, 29–39.

Priest, N.D. (2001) Toxicity of depleted uranium. *Lancet*, 357, 244–246.

Roshchin, A.V., Ordzhonikidze, E.K., and Shalganova, I.V. (1980) Vanadium: Toxicity, metabolism, carrier state. *J Hyg Epidemiol Microbiol Immunol*, 24, 377–383.

Sanchez, D.J., Ortega, A., Domingo, J.L., and Corbella, J. (1991) Developmental toxicity evaluation of orthovanadate in the mouse. *Biol Trace Elem Res*, 30, 219–226.

Scialli, A.R. (1988) Is stress a developmental toxin? *Reprod Toxicol*, 1, 163–171.

Shields, L.M., Wiese, W.H., Skipper, B.J., Charley, B., and Benally, L. (1992) Navajo birth outcomes in the Shiprock uranium mining area. *Health Phys*, 63, 542–551.

Slanina, P., Frech, W., Ekström, L.G., Lööf, L., Slorach, S., and Cedergren, A. (1986) Dietary citric acid enhances absorption of aluminum in antacids. *Clin Chem*, 32, 539–541.

Tannenbaum, A. (1951) *Toxicology of Uranium Compounds*, McGraw-Hill, New York.

Voegtlin, C. and Hodge, H.C. (1953) *Pharmacology and Toxicology of Uranium*, McGraw-Hill, New York.

Weberg, R., Berstad, A., Ladehaug, B., and Thomassen, Y. (1986) Are aluminum containing antacids during pregnancy safe? *Acta Pharmacol Toxicol*, 59, S63–S65.

Wide, M. (1984) Effect of short-term exposure to five industrial metals on the embryonic and fetal development of the mouse. *Environ Res*, 33, 47–53.

Winneke, G., Lilientahl, H., and Kramer, U. (1996) The neurobehavioral toxicology and teratology of lead. *Arch Toxicol*, 18(suppl), 57–70.

Wrenn, M.E., Durbin, P.W., Howard, B., Lipsztein, J., Rundo, J., Still, E.T., and Willis, D.L. (1985) Metabolism of ingested U and Ra. *Health Phys*, 48, 601–633.

Yokel, R.A., Ackrill, P., Burgess, E., Day, J.P., Domingo, J.L., Flaten, T.P., and Savory, A. (1996) The prevention and treatment of aluminum toxicity including chelation therapy. Status and research needs. *J Toxicol Environ Health*, 48, 667–683.

Zaire, R., Notter, M., Riedel, W., and Thiel, E. (1997) Unexpected rates of chromosomal instabilities and alterations of hormone levels in Namibian uranium miners. *Radiat Res*, 147, 579–584.

chapter 5

Intrauterine and Reproductive Toxicity of Nutritionally Essential Metals

Mari S. Golub
Department of Environmental Toxicology
University of California, Davis

Contents

Table 5.1 Nutritionally Essential and Beneficial Elements (metals are in **bold**)

Nutritionally Essential Elements	Elements with Possible Beneficial Effects
Chromium (+3)	Arsenic
Cobalt	Boron
Copper	**Nickel**
Iron	Silicon
Manganese	**Vanadium**
Molybdenum	
Selenium	
Zinc	

Introduction

The Food and Nutrition Board of the National Academy of Sciences reviews nutrition research and recommends values for daily intakes of nutrients. Several elements (Table 5.1) have been designated as "nutritionally essential" or as having "possible beneficial effects" by the Food and Nutrition Board (National Academy of Sciences 2000). Many of these elements are metals, and their toxicity is reviewed as a group in this chapter. An exception is vanadium, a metal with beneficial nutritional properties, which is reviewed in the Chapter 4.

Many essential elements (iron, zinc, manganese, copper, molybdenum) are transition metals and the basis for their essentiality is their role as cofactors in enzymes. Nickel is also thought to modify enzyme activity, but specific nickel-containing enzymes have not been identified. Cobalt is a component of Vitamin B12, while chromium and vanadium are involved in glucose tolerance. It should be noted that developmental and reproductive consequences are associated with *deficiencies* of several of these metals, including in particular deficiencies of zinc, manganese, and copper, which are teratogenic (Keen 1996).

Essential metals are components of many chemicals in commerce and of pharmaceuticals. Only studies of dissociable metal salts are reviewed in this chapter. This review is also limited in that it does not include studies of fertility in nonmammalian species or in production animals (livestock). There is a substantial literature in this area, particularly regarding aquatic species.

Essential metals vary considerably in their tissue concentrations. Iron, zinc, copper, and manganese are found in μM concentrations in human serum, whereas chromium, cobalt, nickel, and molybdenum are found in nM concentrations. Therefore, exposure to environmental media will differentially affect existing body pools of the essential metals at a similar dose. This is an important consideration in comparing the effective toxic doses of the essential metals.

Studies in Humans

Very little information is available on the reproductive consequences of exposure to excess amounts of essential metals in humans. A few studies are available on iron and copper. Overload of iron and copper can occur as a result of use as a supplement or in the hereditary conditions human hemochromatosis and Wilson's Disease.

Fetal loss and malformation were studied in a series of patients with iron overdose in pregnancy. A retrospective study was conducted in London of 49 patients who were overdosed with iron during pregnancy (McElhatton et al. 1991). Thirty-eight of the patients were treated for the overdose with chelators or procedures to empty the stomach. Information on pregnancy outcome was obtained from physicians, but there was no control group. Thirty-seven mothers had normal pregnancy outcomes. There were 3 spontaneous abortions, four elective abortions, and one abortion due to trauma in the case series. Three women delivered early. Five malformations were reported, but the authors felt that the timing of the overdose was not appropriate for a causal relation in three cases. For the other two malformations (talipes and heart murmur with possible ventricular septal defect) a causal relationship was considered possible but unlikely. The authors reviewed eight other case reports from the literature. None of these babies had malformations; however, only one of the poisonings occurred during organogenesis. Five of the eight had normal pregnancy outcomes. The other three cases resulted in maternal death; in two cases, the late-term fetuses survived.

Fetal loss was a prominent outcome in an animal model for iron overdose during pregnancy (Beliles & Palmer 1975). Rabbits and rats received three intravenous injections of iron dextran, either 5, 20, or 50 mg Fe/kg/d for 3 days in late pregnancy. The iron dextran was given by i.v. infusion as in clinical situations. Offspring mortality was increased at birth and for several days after birth in the rabbits. The iron dextran was generally more toxic to the rat dams, making the rat experiment difficult to interpret, but postnatal pup mortality was also seen. Postnatal growth retardation was attributed by the authors to poor maternal condition.

A polymorphism in the Hfe gene can lead to an iron overload syndrome in humans, one symptom of which is hypogonadism, sometimes associated with infertility. The link to iron overload has been established by successful fertility treatment by phlebotomy in individual patients (Oehninger et al. 1998; Cundy et al. 1993). Individuals homozygous or heterozygous for the polymorphism do not generally demonstrate decreased fertility in terms of children per couple, possibly due to the late onset of infertility in the reproductive lifespan (De Braekeleer 1993). Patients with beta-thalassaemia also experience iron overload and impaired fertility. In these patients, sexual maturation is delayed and adults have hypgonadotrophic hypogonadism (Chatterjee & Katz 2001; Chatterjee et al. 1993). Sperm also demonstrate DNA damage in this syndrome (Perera et al. 2002). This is supported by an animal study that found oxidative damage, including DNA damage, and increased sperm with abnormal morphology in mice injected with iron dextran (Doreswamy et al. 2004; Doreswamy & Muralidhara 2005).

There is also a genetic disorder that leads to excess copper accumulation (Wilson's Disease). Case reports have described hypogonadism and impotence in men, and amenorrhea and abortion in women, with untreated, symptomatic Wilson's Disease. In a questionnaire follow-up of treated (n = 15) and untreated (n = 31) female and male (n = 27) patients with Wilson's disease, four women had trouble conceiving and one man was impotent (Tarnacka et al. 2000). There was no control group in the study. Treatment consisted of metal chelation (d-penicillamine) or zinc sulfate, which decreases copper uptake. A similar incidence of spontaneous abortions, preeclampsia, prematurity, and stillbirth was seen in the pregnancies of treated and untreated women. Three infants (4.8% of pregnancies) in the untreated group had congenital heart defect (ventricular septal defect). Heart defects have also been reported in animal teratology studies with copper, which are discussed below. One of the 27 men with Wilson's Disease was self-reported as impotent.

Exposure to several metals discussed in this chapter occurs in connection with welding. Studies of male reproductive toxicity in welders are reviewed in Chapter 9. A study of men exposed to chromium through work in an electroplating factory is available (Li et al. 2001). Semen samples were collected from 21 young exposed men and 22 controls. Blood samples were also collected for testosterone and gonadotropin assays. Although serum chromium concentrations were similar in the two groups, sperm counts and sperm motility were significantly lower in the exposed group, and serum FSH was elevated, suggesting testicular toxicity. Testicular toxicity has also been seen in experimental animals exposed to chromium, as described below.

Studies in Animals

Intrauterine Toxicity

Several research groups have studied a variety of metals in a single paradigm to determine and compare their teratogenic and embryotoxic potential. These approaches include *in vitro* and *in vivo* models in mammalian and nonmammalian systems. The relative potencies of the essential metals in these test systems are summarized in Table 5.2.

Table 5.2 Relative Developmental Toxicity of Essential Metals Based on Dose in Various Teratology Screening Assays (see text for study references)

Assay	Order of Toxicity
Preimplantation mouse embryo *in vitro*	Cr^{+6}, Zn > Mn^{+2}, Mn^{+3}, Cu > Fe^{+2}, Fe^{+3}, Cr^{+3}
Frog embryo (FETAX)	Ni > Cu > Co > Zn
Chick embryo, *in ovo*	Co, Cu > Mo, Mn, Fe
Rodent *in vivo*, i.p. or i.v. injection	Ni > Cu > Zn, Mn >> Co, Mo (no effect)

In Vitro/*Nonmammalian Studies*

Comparison of 12 metals, including iron, manganese, chromium, copper, and zinc, was conducted in preimplantation mouse embryo culture (Hanna et al. 1997). Iron, manganese, and chromium were studied in more than one valence state. A series of 14 media concentrations (range 0 to 200 µM) were used for each agent to determine effective doses. Among the essential elements, Cr^{+6} and Zn^{+2} were effective at lower concentrations than the other essential metals in preventing blastocyst formation and reducing the number of cells in the blastocysts that were formed. Mn^{+2}, Mn^{+3}, and Cu^{+2} were effective at similar concentrations, while iron (Fe^{+2}, Fe^{+3}) and Cr^{+3} were not effective in the concentration range used. The authors pointed out that copper and zinc were effective at concentrations at or below the normal human serum concentrations. Concentrations of these elements in oviductal and uterine fluids are not known. None of the essential metals were as potent as the toxic metals mercury, cadmium, or arsenic in the preimplantation assay; however, their potency was similar to that of lead. The authors concluded that neither essentiality nor redox potential was associated with embryotoxicity across metals in this system.

Metals have been studied for embryotoxicity in two nonmammalian models, chick and frog (Xenopus laevi) embryos. In both models the metals are applied directly to the embryo (by injecting the chick egg or growing the frog embryos in metal-containing media). Sunderman and colleagues studied the developmental toxicity of cobalt, nickel, copper, and zinc in the FETAX (Frog Embryo Teratogenesis Assay: Xenopus) assay in which frog embryos are cultured in medium containing the toxicant to the tadpole stage, and then evaluated either immediately or after metamorphosis for external and internal malformations (Plowman et al. 1994; Plowman et al. 1991). Nickel was the most potent teratogen in this system, producing 55% malformed frogs at 2.3 µmol/L. Copper produced 50% malformations at 2.5 µmol/L, cobalt produced 40% malformation at 25 µmol/L, and zinc produced 50% at 40 µmol/L.

Studies by Gilani and collaborators in chicks included the essential elements chromium (Gilani & Marano 1979), nickel (Gilani & Marano 1980), cobalt, copper, molybdenum, manganese, and iron (Gilani & Alibhai 1990). For the latter five essential elements, at least four doses were tested in order to determine LD50 doses in terms of µg/egg. Endpoints were body weight at the end of organogenesis, gross malformations, and hemorrhage. Cobalt and copper had lower effective doses (on a µg/egg basis) than molybdenum, manganese, and iron, although all the essential metals were less effective than the toxicants arsenic and cadmium. Characteristic and dose-responsive patterns of abnormalities were seen for cobalt and copper.

In Vivo *Mammalian Injection Studies*

A number of studies of metal-induced developmental toxicity in rodents have been conducted using the intraperitoneal and intravenous routes of administration. In particular Ferm and colleagues studied a variety of essential and nonessential metals in the hamster and mouse in order to determine their ability to cause malformations (teratogenicity).

Malformations were also found in fetuses when *copper* (10 µg/kg) and *nickel* (2 to 25 mg/kg) were administered by i.v. injection on gd 8 to hamsters (Ferm 1972; Ferm & Hanlon 1974). The malformations induced by copper included heart defects, and heart defects were also described after i.v. administration of copper to hamsters by Mas et al. (1985), DiCarlo (1980), and DiCarlo (1979). However, *cobalt* acetate, *manganese* chloride (10 to 35 mg/kg) and metallic *molybdenum* (up to 100 mg/kg) were not teratogenic using the i.v. hamster model (Ferm 1972). Since the major interest in these studies was induction of malformation, other developmental toxicity endpoints were not extensively reported, but both maternal death and resorptions occurred at the high doses. *Chromium* (Cr^{+6}, 8 mg/kg) was also teratogenic by the i.v. route of administration in hamsters (Gale & Bunch 1979). Rats treated by the i.v. route with 2 mg Cr/kg on gd 8 did not demonstrate fetal toxic effects (Mason et al. 1989). Single i.p. injections of *zinc* chloride (12.5, 20.5, or 25 mg/kg) on gd 8, 9, 10, or 11 to mice led to skeletal ossification deficits and increased incidence of wavy ribs, but no increase in soft tissue malformations (Chang et al. 1977). The higher doses produced maternal mortality in that study.

The teratogenicity of *nickel* by i.p. injection in mice was confirmed in an extensive study (Lu et al. 1979) that administered 6 doses (1.2 to 6.9 mg Ni/kg) on each of 5 individual days during organogenesis (gd [gestation day] 7 to 11). Pregnancy outcome was determined on gd 18; fetal exams included gross and skeletal abnormalities. Several dams died at the highest dose on gd 7 and 8, and at the three highest doses on gd 9 to 11. Dose-response relationships were seen for implantation, fetal weight, fetal viability, and skeletal abnormalities, and patterns of malformations that varied by day of treatment were reported. The most common abnormalities reported by Lu et al. (1979) in the treated groups were vertebrae and/or rib fusion, cleft palate, open eyelid, clubfoot, and ankylosis. Teratogenic properties of nickel were also studied in rats by i.p. injection (Mas et al. 1985). Nickel chloride was administered at 3 doses (1, 2, and 4 mg/kg) on three different gestation days (gd 8, 12, and 16). These doses were below the LD50, which was estimated at 5 to 6 mg/kg in pregnant rats. No information about maternal condition was given in the report. No fetal toxicity was seen when the rats were dosed at the end of embryogenesis (gd 16). Fetal weights were reduced at the two higher doses (2, 4 mg/kg) on gd 12. In addition, an increased fetal incidence of hemorrhage and hydronephrosis were seen in the groups treated on gd 8 and 12. Additionally, hydrocephalus incidence was increased with gd 12 administration.

As noted above, agents that elicited malformations by parenteral routes in mice and hamsters were not necessarily effective in rats. For instance, nickel administered by the i.m. route on gd 8, or gd 6 to 10 did not produce malformations in rats (Sunderman et al. 1978). However, an increased incidence of dead fetuses and lower fetal weights were seen. The finding of perinatal mortality after gestational nickel injection was more dramatic in a study (Diwan et al. 1992) in which all pups died within a 3-day period after birth, when rats were injected i.p. with nickel on gd 12, 14, 16, and 18.

Manganese injected by the i.p. route to rats at 10 mg/kg had no apparent effect on fetal endpoints (Kimmel et al. 1974). However in mice, a single i.p. injection of 12.5 to 50 mg/kg manganese produced exencephaly, embryonic loss, and fetal growth retardation (Webster & Valois 1987). In another mouse study, manganese was injected subcutaneously at 2 to 16 mg/kg/d (Sanchez et al. 1993). No malformations were reported but reduced fetal weights and morphological abnormalities were seen at the higher doses. Although manganese was not recognized as a teratogen in early studies, more recently, a specific syndrome of skeletal effects has been demonstrated in rats injected i.v. with *manganese* (as manganese chloride) on gd 6 to 17 (Grant et al. 1997; Treinen et al. 1995; Blazak et al. 1996). This extended period of treatment was adopted because studies with a manganese-containing contrast agent (mangafodipir trisodium) indicated a sensitive period from gd 15 to 17 for induction of the skeletal effects, which consisted of abnormal shapes of the long bones, as well as the clavicle and scapula.

A single study of i.v. *iron* administration is available. As described above, this study was intended to model overdose of iron therapy during the third trimester when anemia often develops in pregnant women. Conceptus mortality at birth and in the early postnatal period was the main effect (Beliles & Palmer 1975).

In Vivo *Mammalian Oral Administration Studies*

Some essential trace metals have also been studied for intrauterine toxicity by the oral route of administration, which is more relevant to human exposures. In general, positive findings were reported for nickel, copper, chromium, and cobalt, while manganese and molybdenum were not effective. Iron and zinc have been minimally studied by oral routes for intrauterine toxicity.

Three studies of excess *copper* in pregnancy were available for review only from secondary sources. Keen (1996) described a study in which copper was administered to mice in feed at concentrations of up to 500 µg Cu/g diet. No effects of copper on pregnancy outcome or gross malformation were seen in spontaneously delivered fetuses. Effects on viability were thought to be mediated by reduced dam food intake. When administered to rats in drinking water before and during pregnancy, the main effects of copper were growth retardation in term fetuses and reduced axial and appendicular skeletal ossification (Haddad et al. 1991). The concentration of copper in the water was greater than 1500 mg/L. A third study administered copper sulfate in diet at seven concentrations (26 to 208 mg/kg/d) from one month prior to mating through pregnancy to mice (Leyck 1980, reviewed in (Schardein 1993)). Secondary sources imply that malformations were increased at the higher doses. No studies administering copper by gavage during organogenesis were found. It would be valuable to know if cardiac defects seen in *in vitro* and injection studies also occur after oral administration during organogenesis.

Extensive studies of *chromium* (Cr^{+6}) developmental and reproductive toxicity have been published by the Industrial Toxicology Research Centre in Lucknow, India (see Table 5.3, studies 1 through 3). These three studies administered chromium (Cr^{+6}, potassium chromate) in drinking water to mice during pregnancy. Doses of 250 to 1000 mg/L were used. The authors estimate the daily Cr^{+6} dose at about 65 to 300 mg Cr/kg/d. Dose-dependent accumulation of chromium in maternal serum, the placenta, and the fetus were demonstrated in these studies. One study administered Cr^{+6} in drinking water throughout pregnancy in mice at 250, 500, and 1000 µg/mL (Trivedi et al. 1989). At the lowest dose (250 µg/mL), increased resorptions, postimplantation loss, decreased fetal weight, and decreased ossification were noted, but there was no report of increased malformation. Additionally, at the mid-dose, gross and skeletal malformations were increased. No implantations occurred at the highest dose. In other studies, Cr^{+6} was administered only during organogenesis (gd 6 to 14) (Junaid et al. 1996) or the fetal period (gd 15 to 19) (Junaid et al. 1995) to mice. The doses were 0, 250, 500, or 750 µg/mL. A full spectrum of fetal effects was seen with dosing during organogenesis at the two higher doses, including pre- and postimplantation loss, increased resorptions, decreased litter size, and decreased fetal weight. Malformations seen at the two highest doses were kinky tail and short tail as well as ossification defects. Growth deficits (placental weight, fetal weight, fetal crown-rump) were prominent with the fetal period dosing. Pregnancy weight gain was reduced by 13 and 20%, respectively, at the two higher doses. Effects seen at the lowest dose were on resorptions and litter size with gd 6 to 14 dosing, and fetal weight and length with gd 15 to 19 dosing. These experiments demonstrated intrauterine toxicity of Cr^{+6} at doses at or above about 60 mg/kg/d. Unfortunately, lower doses were not tested.

In an extensive series of experiments, Szakmary and colleagues (2001) studied *cobalt* administered by gavage to pregnant mice, rats, and rabbits. They demonstrated dose-dependent accumulation of cobalt in maternal tissues and distribution to the fetus in rats. Cobalt sulfate (25 to 100 mg/kg) led to greater accumulation than equivalent doses of cobalt chloride and was used in subsequent experiments. Fetal parameters affected in rats, when these doses were administered throughout pregnancy, included retardation in skeletal ossification (25, 50, and 100 mg/kg), and visceral development (50 and 100 mg/kg), and a small but significant increase in total skeletal and visceral malformations (50 and 100 mg/kg). Maternal weight gain was not affected, although food consumption was lower than controls in the highest dose group. In mice treated with 50 mg/kg during organogenesis only, an increase in total malformations and delayed skeletal development were seen, as well as fetal growth retardation. No maternal toxicity was seen in the mice. When doses of 20, 100, or 200 mg/kg were administered to rabbits during organogenesis, the two highest doses proved lethal to the rabbit dams. The 20 mg/kg dose produced 25% mortality and extensive maternal toxicity, so that interpretation of fetal data was limited. In another study (Paternain et al. 1988), no effects on fetuses were found when *cobalt* sulfate was administered by gavage on days 6 to 15 of gestation to rats in a standard teratology design. The doses ranged from 25 to 100 mg/kg and represented one-twentieth to one-fifth the LD50. The dams experienced toxicity as indicated by altered hematology. Food intake and maternal weight gain were reduced at the highest dose, and the amount of cobalt ingested was also reduced at that dose. Thus, high peak exposures of cobalt, similar to those in the Szakmary et al. experiment, were not achieved.

As was the case for the *in vitro* and injection studies described above, *manganese* had little effect on intrauterine toxicity when administered by the oral route in early studies. Manganese was studied by gavage during organogenesis in three species (Food and Drug Laboratories 1973, described in Barlow & Sullivan 1982). The doses and species were 78 mg/kg in rats, 136 mg/kg in hamsters, and 112 mg/kg in rabbits. No adverse effects on the dam or fetus were seen in any of the species. Manganese administered via drinking water at doses up to 20,000 mg/L had no effect on litter size in spontaneously delivered litters. Litter weight was reduced at the highest dose (Kontur & Fechter 1985; Kontur & Fechter 1988).

Summary of Essential Metal Intrauterine Toxicity

Table 5.2 compares the embryotoxicity of essential metals studied within the same system.

At this level of comparison, nickel, cobalt, and copper appear to be more potent than chromium, molybdenum, iron, and manganese, in terms of *in utero* toxicity.

Fertility and Reproductive Toxicity

In studies of fertility, metals are generally administered orally for an extended period of time before, during, and after mating. In most cases, the studies are not designed to detect the specific time or mechanism of action of the metal. Reproductive impairment in males was most frequently reported for the essential metals. Additional studies have specifically examined testicular toxicity. Chromium and cobalt have been most extensively studied for male reproductive toxicity with less information available for copper, molybdenum, and manganese. Information on zinc is available primarily in abstracts. The limited information generated for iron in laboratory animals was discussed previously in this chapter in connection with human studies of iron overload.

Several essential metals have been tested in the dominant dethal assay. In this experimental paradigm, mice are dosed prior to mating with a succession of females over a multiweek period, usually corresponding to the length of the sperm cycle. The main endpoint is postimplantation loss in impregnated females. This test is usually considered a genotoxicity assay because any effect on embryo/fetal death would presumably result from damage to the sperm by the agent. However, the test also provides information on male infertility, based on the number of females impregnated. In addition, male-mediated intrauterine toxicity can be considered an important fertility endpoint. Thus dominant lethal assays are included in the studies reviewed below.

Chromium

Due to its prominence as an environmental toxicant, chromium as potassium dichromate (Cr^{+6}) has been widely studied in standard subchronic, multigeneration, and continuous breeding toxicology protocols. Very limited information is available for Cr^{+3}, but some studies using both Cr^{+3} and Cr^{+6} found similar results (Elbetieha & Al-Hamood 1997; Bataineh et al. 1997; Zahid et al. 1990). Although Cr^{+6} is not an essential nutrient, studies of Cr^{+6} are included here under the general review of chromium.

Spermatogenesis was a focus of these Cr^{+6} studies. Concern about chromium effects on spermatogenesis were raised in a study that found decreased sperm counts in mice fed either Cr^{+6} (potassium dichromate) or Cr^{+3} (chromium sulfate) in diet for 5 weeks (Zahid et al. 1990). The doses were 100, 200, or 400 µg Cr^{+6}/g diet, or about 15, 30, and 60 mg/kg bw/day. A later study in rats (Wistar) found decreased sperm counts when Cr+6 (chromium trioxide, 10 or 20 mg /kg) was administered by feeding for 6 days and the sperm counts were conducted 6 weeks later (Li et al. 2001). Abnormal sperm morphology was also increased in the treated groups. However, subchronic (9-week) studies conducted by the National Toxicology Program (NTP 2002) using the same strain of mice (BALB/c) but a different strain of rats (Sprague Dawley) as the Zahid and Li studies, did not find effects on sperm count after administration of a somewhat lower dose range (0, 15, 50, 100, or 400 µg Cr^{+6}/g diet) for 3, 6, or 9 weeks (NTP 1996, 1997).

As part of their series of studies of Cr^{+6} developmental and reproductive toxicity (Table 5.3, studies 8 and 9) investigators in Lucknow, India, looked at the effect of injected Cr^{+6} on sperm and testes. Using i.p. dosing, no effects on sperm parameters were found when 2 mg Cr^{+6}/kg/d was injected for 15 days in adult rats (Murthy et al. 1991). However, histology revealed some damage to Sertoli cell junctions and to the spermatid midpiece. Effects on immature rats were also studied using doses of 1, 2, or 3 mg/kg administered i.p. (Saxena et al. 1990). Endpoints affected were testicular enzymes and some aspects of testicular pathology.

Thus of five studies investigating Cr^{+6} effects on testes and sperm parameters, only the NTP studies failed to find some evidence of testicular effects. This discrepancy could be due to differences between studies in detail of compound administration, sperm evaluation, test species, diet and age, and so forth.

Fertility has been evaluated specifically in Cr^{+6}-treated male rats in one study (Bataineh et al. 1997). This study additionally evaluated Cr^{+3} (chromium chloride). Both compounds were administered in drinking water at a

Table 5.3 Developmental and Reproductive Toxicity Studies Involving Chromium

Study First Author, Year	Species	Doses	Dosing Period	Evaluation
1. Trivedi 1989	mouse	0, 250, 500, 1000 µg/mL drinking water	Gd 1–19	Fetal exam gd 19
2. Junaid 1995	mouse	0, 250, 500, 750 µg/mL drinking water	Gd 15–19	Fetal exam gd 19
3. Junaid 1996a	mouse	0, 250, 500, 750 µg/mL drinking water	Gd 6–14	Fetal exam gd 19
4. Junaid 1996b	mouse	0, 250, 500, 750 µg/mL drinking water	20 days premating	Estrous cycle Mating index Fertility index Fetal exam gd 19
5. Kanojia et al. 1996	rat	0, 250, 500, 750 µg/mL drinking water	20 days premating	Estrous cycle Mating index Fertility index Fetal exam gd 19
6. Kanojia et al. 1998	rat	0, 250, 500, 750 µg/mL drinking water	90 days premating	Estrous cycle Mating index Fertility index Fetal exam gd 19
7. Murthy 1996	mouse	0, 250, 500, 750 µg/mL .05, .5, 5 µg/mL drinking water	20 days 90 days premating	Ovarian histology Ovarian ultrastructure
8. Saxena 1990	rat immature	1, 2, 3 mg Cr/kg i.p.	Weaning–pnd 55 Weaning–pnd 90	Sperm evaluation Testes pathology Testes enzymes
9. Murthy 1991	rat adult	2 mg Cr/kg i.p.	15 days	Sperm evaluation Testes pathology

concentration of 1000 mg/L for 12 weeks. The chromium-treated males in both groups weighed about 25% less than the controls by the end of the study. Mating and aggressive behaviors were evaluated prior to a fertility trial in which the males were housed with two females for ten days. Both forms of chromium reduced the number of mounts, the number of males ejaculating and the postejaculatory interval. Only Cr^{+6} increased the time to ejaculation. Intromissions were not affected. Aggressive behaviors elicited by encounter with a strange male were lower in both chromium-treated groups. Pregnancy rates and number of implantations per female did not differ from the control, but resorptions and dead fetuses were increased in females mated with the chromium-exposed males.

In addition to this evaluation of chromium, the same authors (Bataineh et al. 1997) evaluated copper and manganese in the same paradigm, using 1000 mg/L drinking water concentrations. With respect to mating behavior, effects of manganese were found on ejaculatory latency and postejaculatory interval, while copper affected a broad range of variables including time to mount, intromission latency, number of intromissions, ejaculatory latency, and postejaculatory interval. Both copper and manganese treatment reduced aggression. Neither agent reduced fertility in the small groups of males tested (n = 9 to 12/group), but manganese treatment led to a higher incidence of resorptions/implantations in pregnancies from the treated males. Other fertility studies are available for Cr^{+6} using continuous breeding designs, in which both males and females are dosed. These studies used potassium dichromate (Cr^{+6}) in the diet of BALB/c mice or SD rats at concentrations of 0, 100, 200, and 400 µg/g and found no effects on fertility in either the F0 or the F1 generation (NTP 1996; NTP 1997).

As regards female reproductive toxicity of Cr^{+6}, there are also several studies addressing female endpoints from the Lucknow group. Study designs are outlined in Table 5.3 (studies 4 through 7).

Two studies administered the same drinking water concentrations of Cr^{+6} to mice (Junaid et al. 1996a) and rats (Kanojia et al. 1996) for a 20-day period prior to mating. This dosing period was intended to cover a complete cycle of ovarian follicle development. The group size was larger for mice (n = 15) than rats (n = 10). After dosing was completed, the animals were mated and fetal exams were conducted at the end of pregnancy. The drinking water concentrations were more toxic to mice than rats, producing 20% mortality in the mice at the highest concentration. However, differential water intake of the two species needs to be taken into account; the mg/kg/d doses of chromium were about threefold higher for the mice than the rats. Reduced pregnancy weight gain was seen at the two higher doses in rats and in all three doses in mice. Decreased fetal viability, and decreased fetal weight and length were seen at the low and mid dose in mice (the high-dose dams were not fertile). Fetal viability effects were also seen in the rat study at the mid and high doses, but fetal growth was not significantly affected. A higher incidence of subdural hemorrhage, tail abnormalities (short, kinky), and ossification delays were seen in rats at the high dose, and in mice at the mid dose. Thus, similar effects of premating Cr^{+6} exposure were seen in mice and rats.

The rat study additionally included evaluation of length of estrous cycles, and mating and fertility indices. All three parameters demonstrated a dose-related pattern (longer estrous cycles, reduced fertility) with significant differences from control at the highest dose. The rat was also used in a study with a longer period of pregestational chromium exposure (90 days) (Kanojia et al. 1998). The drinking water concentrations were the same as previous studies (0, 250, 500, and 750 mg/L). The two higher doses proved fatal to 15 and 10% of the treated females. The pattern of fetal effects and the effective dose levels were similar to those of the shorter 20-day period of exposure. However, the rats with the longer exposure period were anovulatory (persistent diestrous) at the conclusion of dosing so that pregnancies were not established until 15 to 20 days later, allowing for elimination of chromium and recovery of function. Nonetheless, fetal chromium concentrations were over tenfold higher after the longer versus the shorter pregestational exposure.

A final study in this series on female reproductive effects of Cr^{+6} (Murthy et al. 1996) looked in detail at induced ovulation and ovarian histopathology. This study was conducted in mice. Drinking water concentrations of 0, 250, 500, and 750 mg/L for 20 days were supplemented with a lower concentration series of 0.05, .5, and 5 mg/L for 90 days. The lower dose range reflected the World Health Organization–recommended drinking water limit of .05 mg/L. The mice in the low-dose range were used only for ultrastructure evaluation of the ovary by EM (electron microscope). As expected from the infertility and reduced number of corpora lutea in the previous study, the ovaries of the 750 mg/L mice showed follicular atresia and stromal damage. At the two highest doses, reduced ovulation was seen in response to gonadotropin. Some indication of ovarian damage was seen by EM in the mice treated with 5 mg/L Cr^{+6} in drinking water. There was damage to the cell membrane of ovarian follicles, extracellular fluid accumulation, and changes in mitochondrial structure.

Given the adverse reproductive effects of chromium on adults, it can be suggested that chromium would also influence reproductive maturation. One study addressed this issue by exposing developing mice to Cr^{+6} or Cr^{+3} via treatment of their dams during gestation and lactation. The chromium was administered in drinking water at a concentration of 1000 mg/L for both agents (potassium dichromate and chromium chloride) beginning on gd 12. Fertility of both male and female offspring was evaluated with mating trials on postnatal days (pnd) 60. Although fertility of the male offspring (10/group) was not affected by the chromium treatments, the relative reproductive organ weights (testes, seminal vesicle, preputial gland) were markedly reduced in the Cr^{+3} group. In female offspring, group sizes were n = 12 for controls control, n = 16 for those treated with Cr^{+3}, and n = 22 for those treated with Cr^{+6}. Vaginal opening was delayed by both Cr^{+3} and Cr^{+6} treatments, and the number of females becoming pregnant was lower in both the

Cr^{+3} and Cr^{+6} groups. Implantations and viable fetuses were lower only in the pregnant females of the Cr^{+6} group.

Cobalt

Male reproductive toxicity has been a focus of *cobalt* toxicity studies in animals. Early studies involved injection of cobalt for 30 days and found testicular pathology (Hoey 1966). Later, chronic oral exposure to cobalt was found to lead to testicular toxicity (atrophy, reduced sperm, epithelial degeneration). Subsequently, Pedigo and collaborators (Pedigo et al. 1988; Pedigo & Vernon 1993) provided a particularly thorough examination of male fertility for *cobalt* chloride. Male mice (n = 5/group) were exposed to cobalt in drinking water at concentrations of 100, 200, or 400 and were evaluated after 12 weeks of treatment for fertility, sperm parameters, reproductive organ weights, and serum testosterone and gonadotropins. Body weights were reduced about 10% relative to controls in the high-dose group, while relative testes weight and epididymal sperm counts were significantly reduced in all dose groups by the end of treatment in a dose-dependent pattern. The high-dose group also demonstrated reduced sperm motility and severely impaired fertility as measured by the number of fertilized ova in the oviducts of females mated two days after mating. Elevated serum testosterone was found in all three dosage groups with some (nonsignificant) indication of elevation of gonadotropins.

Pedigo et al. further examined the cobalt effects at the high dose in an experiment that included pregnancy outcome. When male mice were injected with 200 μmol Co/Kg i.p. for 3 days prior to the 7-week mating period, in a classic dominant lethal paradigm, there was a nonsignificant increase in preimplantation loss (number of implantations/number of eggs ovulated as reflected in ovarian corpora lutea). Thus acute exposure did not produce clear male reproductive toxicity.

However, when mice were exposed to cobalt continuously via drinking water dosing (400 μg Co/mL) for 10 weeks and then mated for 2 weeks, significant differences from controls were seen in the numbers of females impregnated, the number of implantations per pregnant female, and the preimplantation loss per pregnant female. When sperm from chronically treated mice were studied using automated technology, deficits in sperm numbers and motility were found. DNA damage was not studied; however, another experiment found no abnormalities in embryos collected and cultured from females mated with males treated over a 9-week period. A final assessment studied fertility by mating males with superovulated females and counting the number of fertilized ova in the oviducts 2 days later. In matings conducted immediately after the 2-week period when term pregnancies were evaluated, matings of treated males produced very few or no fertilized ova, as seen in the previous study. After an 8-week recovery period, fertility and sperm parameters recovered, although testes weight remained lower than controls.

Copper

Copper has a unique status as an agent that causes infertility because of its therapeutic use as a contraceptive in the copper intrauterine device (IUD). In its role as a contraceptive, copper has been thought to act directly on sperm, on conceptus viability and/or on the implantation process with the net result of failure of implantation of a viable embryo. However, none of these actions has been established as the basis of contraception (Ortiz et al. 1996). Fertility is also impaired by intrauterine copper in animals, and local application of copper to the testes (intrascrotal) is also effective (Chinoy & Sanjeevan 1980).

In vivo studies of orally administered copper on fertility are rare. Subchronic studies administering copper (as copper sulfate) in drinking water to mice and rats did not find effects on estrous cycling or sperm parameters after 90-day exposures at doses of 300 or 1000 µg/mL (Hebert et al. 1993). Higher doses proved fatal to the rats and mice. As mentioned earlier, copper was one of three metals evaluated for male mating behavior, aggressive behavior, and fertility using a1000 mg/L drinking water concentration (Bataineh et al. 1998). Copper affected a broad range of variables including time to mount, intromission latency, number of intromissions, ejaculatory latency, and postejaculatory interval, and also reduced aggression. Fertility was not reduced, although the incidence of resorptions/implantations in pregnancies from the treated males was increased.

Manganese

Studies discussed above indicated a low potential for manganese to produce intrauterine toxicity in rodent studies. One study in mice also indicated a low potential for reproductive toxicity due to manganese exposure (Elbetieha & Al-Hamood 1997; Bataineh et al. 1998). Manganese chloride was given in drinking water at concentrations of 1000, 2000, 4000, and 8000 mg/L for 12 weeks to male and female mice prior to mating with unexposed partners. Water intake was reduced as manganese chloride concentrations increased. However, effects on fertility were seen only at the highest concentration, 8000 mg/L in both males and females. At that dose, the number of pregnant females out of the number of females mated was decreased in males. In females, fertility indices were not significantly reduced, but the number of implantations and viable fetuses per pregnancy were reduced.

The same authors (Bataineh et al. 1998) evaluated manganese for effects on mating and aggressive behavior in rats using 1000 mg/L drinking water concentrations. With respect to mating behavior, effects of manganese were found on ejaculatory latency and postejaculatory interval. Aggressive behaviors were also reduced in the manganese-treated rats. In agreement with the previous study, fertility was not affected.

Sperm parameters were not measured in the drinking water studies. However, in a gavage study using manganese acetate, testicular sperm counts and sperm motility were reduced in a dose-dependent fashion at doses of 7.5, 25, and 30 mg/kg/day for 43 days (Ponnapakkam et al. 2003). Fertility, assessed only at the highest dose, was not affected. There were no effects of the manganese treatments on reproductive organ weights and testicular pathology.

Molybdenum

Some evidence of male infertility was also obtained for *molybdenum* in a single-generation study. Diets containing less than 1 µg Mo/g were supplemented with 20, 80, or 140 µg Mo/g and fed to small groups of male and female rats (four or eight per group) beginning at weaning, for 11 weeks prior to mating (Jeter & Davis 1954). Growth retardation appeared in the lowest dose group, and weight gain was less than one-half that of controls in the highest dosage group. Six of eight males receiving the 80 or 140 µg Mo/g diets were infertile and demonstrated degeneration of the seminiferous tubules at necropsy. The high dietary molybdenum did not apparently affect fertility and pregnancy in females as judged from a small number of litters that were produced at the two highest doses (seven litters), but maternal weights and pups' weight gain were lower during lactation. Because it is known that molybdenum toxicity is counteracted by copper in the diet, the authors studied molybdenum in a high-copper diet in a small number of rats (n = 4) and found normal fertility. In a three-generation study, molybdenum was given to mice and rats in drinking water at a dose of 10 mg Mo/L (Schroeder & Mitchener 1971). The dose was selected based on previous studies showing absence of chronic toxicity. Treatment commenced at weaning of the F0 generation. Elevated mortality relative to controls was seen in the first two generations, and the third generation failed to generate a fourth generation due to mortality, few litters, and small litter sizes. Because the molybdenum drinking water was continually present, it is not clear whether the mortality and fertility effects were mediated by the reproductive system, or whether males or females were primarily affected.

Very recently, effects on spermatogenesis (sperm counts, sperm motility, and sperm morphology) were demonstrated in rats given tetra thiomolybdate by gavage for 8 weeks at a dose of 12 mg/kg/day (Lyubimov et al. 2004). This agent was developed as a copper chelator to be used in cancer chemotherapy. In addition to effects on spermatogenesis, the same dose produced reduced weight gain, anemia, and copper deficiency. Effects on spermatogenesis were prevented by copper supplementation. No fertility data were generated in these treatment groups; a shorter duration treatment (2 weeks) did not affect fertility when both males and females were treated. Treatment during early pregnancy (gd 0 to 6) had no apparent effect on pregnancy outcome.

Taken together these studies indicate that excess molybdenum can affect testicular function via induction of copper deficiency. However, a dominant lethal effect of molybdenum has also been identified at i.p. doses of 200 and 400 mg/kg, suggesting DNA damage as a mediating event in molybdenum-induced male fertility effects.

Nickel

Early multigeneration studies available for nickel reported effects on perinatal mortality, in particular, increased numbers of dead pups at birth. Similar results have been found in other recent studies with nickel administered orally, usually in drinking water.

Schroeder and Mitchener administered nickel (salt not stated) in drinking water at a concentration of 5 mg Ni/L. All generations of the Schroeder and Mitchener study (Schroeder & Mitchener 1971) had a higher rate of perinatal mortality in the treated group. Lower birth weights (runting) and weaning weights were also associated with treatment in these studies. In their more recent study of nickel effects on female reproduction, Smith et al. (1993) treated female rats for 11 weeks prior to two successive mating, pregnancy, and lactation cycles. Nickel chloride was administered in drinking water at a concentration of 0, 10, 50, or 250 mg Ni/L. No effects on fertility were seen. The number of pups dead at birth was increased at the highest dose for the first pregnancy and for all three doses for the second pregnancy. Perinatal mortality was also identified as a consequence of nickel administration in drinking water when either female rats or both parents were treated (Al-Hamood et al. 1998; Kakela et al. 1999). These authors were able to counteract the effects of nickel by simultaneous administration of selenium.

The recent one- and two-generation studies administered nickel sulfate by gavage (Springborn 2000a, b, described in OEHHA 2001). The single-generation study administered doses in a wider range (10 to 75 mg/kg/d) in order to identify appropriate doses for the two-generation study, which were selected as 0, 1, 2.5, 5, and 10 mg/kg/d. Both male and female parents were dosed. The two-generation study included a full range of endpoints including, fertility, estrous cycles, and sperm evaluations, but found no effects of the nickel treatment. However, the preliminary single-generation study using higher doses found increased postimplantation loss at the 30, 50, and 75 mg/kg/d doses. At the highest dose (75 mg/kg/day) litter size was smaller, and the number of dead pups at birth was higher than in controls. The highest drinking water concentrations in the Smith et al. study would be estimated to result in daily nickel intakes of 20 to 50 mg/kg/d depending on the stage of gestation/lactation. These doses would be more comparable to the single-generation gavage study than the full two-generation gavage study.

Increased perinatal mortality in these chronic oral dosing studies is supported by the findings of increased dead fetuses in studies that injected nickel during pregnancy (Diwan et al. 1992; Sunderman et al. 1978). These studies were reviewed in more detail in the previous section on intrauterine toxicity of nickel.

With respect to male reproductive toxicity, a dominant lethal effect was identified in mice injected i.p. with 56 mg/kg nickel nitrate (Jacquet & Mayence 1982). Several studies additionally reported damage to testes after nickel administration to rats by injection (Mathur et al. 1977; Hoey 1966). A recent paper (Doreswamy & Muralidhara 2005) examined testicular toxicity of nickel chloride injected i.p. in mice at doses of 1.25, 2.5, and 5.0 µmol/ kg/d in some detail. With repeated dosing, an increased incidence of morphologically abnormal sperm was seen, although sperm counts were not affected. At the 2.5 µmol/kg/d dose for 5 days, dominant lethal effects (postimplantation loss) were reported. The authors hypothesized that nickel damages testes through oxidative effects. This was supported by findings of enhanced lipid peroxidation, single-strand DNA breaks, and apoptosis in testes. One study using the oral route of administration (nickel in drinking water) also reported testicular pathology, decreased testicular sperm numbers, and a lower pregnancy rate in rats (Kakela et al. 1999).

Zinc

Zinc reproductive toxicity information has been presented primarily in a series of abstracts concerning a single-generation study in which zinc was given to rats by gavage before mating and during pregnancy and lactation. The doses were 0, 7.5, 15, and 30 mg/kg/d. Zinc effects were described on implantations (decreased), with trends toward decreased viability as reflected in the live birth and weaning indices in the treated groups (Graham et al. 2003). The zinc treatment apparently accelerated maturation of offspring as reflected in morphology (incisor eruption, eye opening, vaginal opening) (Johnson et al. 2003). A similar study using CD-1 mice, also reported as an abstract, used doses of 0, .78, 1.56, and 3.125 mg/kg/d, and also reported adverse effects on implantation and viability (Ogden et al. 2002). No effects on fertility were mentioned for either the rat or mouse study. Another study available in English as an abstract reported that sperm count, motility, and fertility were improved in rats given 12 or 120 mg/kg zinc, but impaired at 240 mg/kg given for 60 days (Wei et al. 2003). The abstract reported no effects on offspring. Full publication or translation of these studies in the future could provide needed information on the reproductive consequences of excess exposure to zinc.

Mechanism Considerations

Mechanisms of intrauterine and reproductive toxicity of essential metals have not been determined. Physical properties of the metals (size, charge) do not seem relevant. Biological actions mentioned or explored by study authors include anemia for cobalt (Paternain et al. 1988), effects on calcium channels for cobalt (Pedigo et al. 1988), glucose regulation for nickel (Mas et al. 1985), zinc deficiency for chromium (Li et al. 2001), hypothalamic-pituitary axis dysfunction for manganese (Elbetieha & Al-Hamood 1997) and copper deficiency for molybdenum (Jeter & Davis 1954; Lyubimov et al. 2004). It is interesting that iron and zinc have limited intrauterine toxicity when administered *in vivo* possibly due to regulatory systems for controlling absorption and distribution of these elements, which are found at relatively high concentrations in food and in the human body. The contribution of maternal toxicity to metal intrauterine toxicity is not known, but maternal–fetal distribution of most of the essential metals was demonstrated in connection with teratology studies. Metals have also been shown to be differentially distributed to the testes, perhaps in connection with acquisition of iron and zinc, which are important for testicular function (Hidiroglou & Knipfel 1984; Sylvester & Griswold 1994). Contemporary research on metal binding and distribution proteins and metal transporters may help elucidate the mechanisms of action of metal reproductive toxicants that are also essential nutrients.

References

Al-Hamood, M.H., Elbetieha, A., and Bataineh, H. (1998) Sexual maturation and fertility of male and female mice exposed prenatally and postnatally to trivalent and hexavalent chromium compounds. *Reprod Fertil Dev*, 10, 179–83.

Barlow, S.M. and Sullivan, F.M. (1982). *Reproductive Hazards of Industrial Chemicals*. Academic Press, New York.

Bataineh, H., al-Hamood, M.H., Elbetieha, A., and Bani Hani, I. (1997) Effect of long-term ingestion of chromium compounds on aggression, sex behavior and fertility in adult male rat. *Drug Chem Toxicol*, 20, 133–49.

Bataineh, H., Al-Hamood, M.H., and Elbetieha, A.M. (1998) Assessment of aggression, sexual behavior and fertility in adult male rat following long-term ingestion of four industrial metals salts. *Hum Exp Toxicol*, 17, 570–6.

Beliles, R.P. and Palmer, A.K. (1975) The effect of massive transplacental iron loading. *Toxicology*, 5, 147–58.

Blazak, W.F., Brown, G.L., Gray, T.J., Treinen, K.A., and Denny, K.H. (1996) Developmental toxicity study of mangafodipir trisodium injection (MnDPDP) in New Zealand white rabbits. *Fundam Appl Toxicol*, 33, 11–5.

Chang, C.H., Mann, D.E.J., and Gautieri, R.F. (1977) Teratogenicity of zinc chloride, 1,10–phenanthroline, and a zinc-1,10-phenanthroline complex in mice. *J Pharm Sci*, 66, 1755–8.

Chatterjee, R. and Katz, M. (2001) Evaluation of gonadotrophin insufficiency in thalassemic boys with pubertal failure: spontaneous versus provocative test. *J Pediatr Endocrinol Metab*, 14, 301–12.

Chatterjee, R., Katz, M., Cox, T., and Bantock, H. (1993) Evaluation of growth hormone in thalassaemic boys with failed puberty: spontaneous versus provocative test. *Eur J Pediatr*, 152, 721–6.

Chinoy, N.J. and Sanjeevan, A.G. (1980) Effects of an intra-epididymal and intra-scrotal copper device on rat spermatozoa. *Int J Androl*, 3, 719–37.

Cundy, T., Butler, J., Bomford, A., and Williams, R. (1993) Reversibility of hypogonadotrophic hypogonadism associated with genetic haemochromatosis. *Clin Endocrinol (Oxf)*, 38, 617–20.

De Braekeleer, M. (1993) A prevalence and fertility study of haemochromatosis in Saguenay-Lac-Saint-Jean. *Ann Hum Biol*, 20, 501–5.

DiCarlo, F.J., Jr. (1979) Copper-induced heart malformations in hamsters. *Experientia*, 35, 827–8.

DiCarlo, F.J., Jr. (1980) Syndromes of cardiovascular malformations induced by copper citrate in hamsters. *Teratology*, 21, 89–101.

Diwan, B.A., Kasprzak, K.S., and Rice, J.M. (1992) Transplacental carcinogenic effects of nickel(II) acetate in the renal cortex, renal pelvis and adenohypophysis in F344/NCr rats. *Carcinogenesis*, 13, 1351–7.

Doreswamy, K. and Muralidhara (2005) Genotoxic consequences associated with oxidative damage in testis of mice subjected to iron intoxication. *Toxicology*, 206, 169–78.

Doreswamy, K., Shrilatha, B., Rajeshkumar, T., and Muralidhara (2004) Nickel-induced oxidative stress in testis of mice: evidence of DNA damage and genotoxic effects. *J Androl*, 25, 996–1003.

Elbetieha, A. and Al-Hamood, M.H. (1997) Long-term exposure of male and female mice to trivalent and hexavalent chromium compounds: effect on fertility. *Toxicology*, 116, 39–47.

Ferm, V.H. (1972) The teratogenic effects of metals on mammalian embryos. *Adv Teratol*, 5, 51–75.

Ferm, V.H. and Hanlon, D.P. (1974) Toxicity of copper salts in hamster embryonic development. *Biol Reprod*, 11, 97–101.

Gale, T.F. and Bunch, J.D., III (1979) The effect of the time of administration of chromium trioxide on the embryotoxic response in hamsters. *Teratology*, 19, 81–6.

Gilani, S.H. and Alibhai, Y. (1990) Teratogenicity of metals to chick embryos. *J Toxicol Environ Health*, 30, 23–31.

Gilani, S.H. and Marano, M. (1979) Chromium poisoning and chick embryogenesis. *Environ Res*, 19, 427–31.

Gilani, S.H. and Marano, M. (1980) Congenital abnormalities in nickel poisoning in chick embryos. *Arch Environ Contam Toxicol*, 9, 17–22.

Graham, T., Ogden, L., Atkinson, A., Johnson, F., Hammersley, M., Wilson, L., DeJan, B., and Knight, Q. (2003) Reproductive effects of zinc chloride in Sprague-Dawley rats. *Toxicologist*, 72, 76.

Grant, D., Blazak, W.F., and Brown, G.L. (1997) The reproductive toxicology of intravenously administered MnDPDP in the rat and rabbit. *Acta Radiol*, 38, 759–69.

Haddad, D.S., al-Alousi, L.A., and Kantarjian, A.H. (1991) The effect of copper loading on pregnant rats and their offspring. *Funct Dev Morphol*, 1, 17–22.

Hanna, L.A., Peters, J.M., Wiley, L.M., Clegg, M.S., and Keen, C.L. (1997) Comparative effects of essential and nonessential metals on preimplantation mouse embryo development in vitro. *Toxicology*, 116, 123–31.

Hebert, C.D., Elwell, M.R., Travlos, G.S., Fitz, C.J., and Bucher, J.R. (1993) Subchronic toxicity of cupric sulfate administered in drinking water and feed to rats and mice. *Fundam Appl Toxicol*, 21, 461–75.

Hidiroglou, M. and Knipfel, J.E. (1984) Zinc in mammalian sperm: a review. *J Dairy Sci*, 67, 1147–56.

Hoey, M.J. (1966) The effects of metallic salts on the histology and functioning of the rat testis. *J Reprod Fertil*, 12, 461–72.

Jacquet, P. and Mayence, A. (1982) Application of the in vitro embryo culture to the study of the mutagenic effects of nickel in male germ cells. *Toxicol Lett*, 11, 193–7.

Jeter, M.A. and Davis, G.K. (1954) The effect of dietary molybdenum upon growth, hemoglobin, reproduction and lactation of rats. *J Nutr*, 54, 215–20.

Johnson, F., Ogden, L., Graham, T., Thomas, T., Gilbreath, E., Hammersley, M., Wilson, L., Knight, Q., and DeJan, B. (2003) Developmental effects of zinc chloride in rats. *Toxicologist*, 72, 75.

Junaid, M., Murthy, R.C., and Saxena, D.K. (1995) Chromium fetotoxicity in mice during late pregnancy. *Vet Hum Toxicol*, 37, 320–3.

Junaid, M., Murthy, R.C., and Saxena, D.K. (1996a) Embryo- and fetotoxicity of chromium in pregestationally exposed mice. *Bull Environ Contam Toxicol*, 57, 327–34.

Junaid, M., Murthy, R.C., and Saxena, D.K. (1996b) Embryotoxicity of orally administered chromium in mice: exposure during the period of organogenesis. *Toxicol Lett*, 84, 143–8.

Kakela, R., Kakela, A., and Hyvarinen, H. (1999) Effects of nickel chloride on reproduction of the rat and possible antagonistic role of selenium. *Comp Biochem Physiol C Pharmacol Toxicol Endocrinol*, 123, 27–37.

Kanojia, R.K., Junaid, M., and Murthy, R.C. (1996) Chromium induced teratogenicity in female rat. *Toxicol Lett*, 89, 207–13.

Kanojia, R.K., Junaid, M., and Murthy, R.C. (1998) Embryo and fetotoxicity of hexavalent chromium: a long-term study. *Toxicol Lett*, 95, 165–72.

Keen, C. (1996). Teratogenic effects of essential trace metals: deficiencies and excesses, in *Toxicology of Metals*, pp. 977–1001, Chang, L., ed., CRC Lewis Publishers, Boca Raton, FL.

Kimmel, C.A., Butcher, R.E., Vorhees, C.V., and Schumacher, H.J. (1974) Metal-salt potentiation of salicylate-induced teratogenesis and behavioral changes in rats. *Teratology*, 10, 293–300.

Kontur, P.J. and Fechter, L.D. (1985) Brain manganese, catecholamine turnover, and the development of startle in rats prenatally exposed to manganese. *Teratology*, 32, 1–11.

Kontur, P.J. and Fechter, L.D. (1988) Brain regional manganese levels and monoamine metabolism in manganese-treated neonatal rats. *Neurotoxicol Teratol*, 10, 295–303.

Li, H., Chen, Q., Li, S., Yao, W., Li, L., Shi, X., Wang, L., Castranova, V., Vallyathan, V., Ernst, E., and Chen, C. (2001) Effect of Cr(VI) exposure on sperm quality: human and animal studies. *Ann Occup Hyg*, 45, 505–11.

Lu, C.C., Matsumoto, N., and Iijima, S. (1979) Teratogenic effects of nickel chloride on embryonic mice and its transfer to embryonic mice. *Teratology*, 19, 137–42.

Lyubimov, A.V., Smith, J.A., Rousselle, S.D., Mercieca, M.D., Tomaszewski, J.E., Smith, A.C., and Levine, B.S. (2004) The effects of tetrathiomolybdate (TTM, NSC-714598) and copper supplementation on fertility and early embryonic development in rats. *Reprod Toxicol*, 19, 223–33.

Mas, A., Holt, D., and Webb, M. (1985) The acute toxicity and teratogenicity of nickel in pregnant rats. *Toxicology*, 35, 47–57.

Mason, R.W., Edwards, I.R., and Fisher, L.C. (1989) Teratogenicity of combinations of sodium dichromate, sodium arsenate and copper sulphate in the rat. *Comp Biochem Physiol C*, 93, 407–11.

Mathur, A.K., Datta, K.K., Tandon, S.K., and Dikshith, T.S. (1977) Effect of nickel sulphate on male rats. *Bull Environ Contam Toxicol*, 17, 241–8.

McElhatton, P.R., Roberts, J.C., and Sullivan, F.M. (1991) The consequences of iron overdose and its treatment with desferrioxamine in pregnancy. *Hum Exp Toxicol*, 10, 251–9.

Murthy, R.C., Junaid, M., and Saxena, D.K. (1996) Ovarian dysfunction in mice following chromium (VI) exposure. *Toxicol Lett*, 89, 147–54.

Murthy, R.C., Saxena, D.K., Gupta, S.K., and Chandra, S.V. (1991) Ultrastructural observations in testicular tissue of chromium-treated rats. *Reprod Toxicol*, 5, 443–7.

National Academy of Sciences. (2000) Dietary reference intakes for vitamin A, vitamin K, arsenic, boron, chromium, copper, iodine, iron, manganese, molybdenum, nickel, silica, vanadium and zinc. Food and Nutrition Board, National Academy Press, Washington, D.C.

National Toxicology Program (NTP). (1996) Final report on the reproductive toxicity of potassium dichromate (hexavalent) (CAS #778-50=9) administered in diet to SD rats. Research Triangle Park, North Carolina.

National Toxicology Program (NTP). (1997) Final report on the reproductive toxicity of potassium dichromate (hexavalent) (CAS #778-50-9) administered in diet to BALB/c mice. Research Triangle Park, North Carolina.

National Toxicology Program (NTP). (2002) Preliminary 13-Week Study Results. Research Triangle Park, North Carolina.

Office of Environmental Health Hazard Assessment (OEHHA). (2001) *Nickel in drinking water: California Public Health Goal.* California Environmental Protection Agency, Sacramento, CA.

Oehninger, S., Pike, I., and Slotnick, N. (1998) Hemochromatosis and male infertility. *Obstet Gynecol*, 92, 652–3.

Ogden, L., Graham, T., Mahboob, M., Atkinson, A., Hammersley, M., and Sarkar, N. (2002) Effects of zinc chloride on reproductive parameters of cd-1 mice. *Toxicologist*, 66, 33.

Ortiz, M.E., Croxatto, H.B., and Bardin, C.W. (1996) Mechanisms of action of intrauterine devices. *Obstet Gynecol Surv*, 51, S42–51.

Paternain, J.L., Domingo, J.L., and Corbella, J. (1988) Developmental toxicity of cobalt in the rat. *J Toxicol Environ Health*, 24, 193–200.

Pedigo, N.G., George, W.J., and Anderson, M.B. (1988) Effects of acute and chronic exposure to cobalt on male reproduction in mice. *Reprod Toxicol*, 2, 45–53.

Pedigo, N.G. and Vernon, M.W. (1993) Embryonic losses after 10-week administration of cobalt to male mice. *Reprod Toxicol*, 7, 111–6.

Perera, D., Pizzey, A., Campbell, A., Katz, M., Porter, J., Petrou, M., Irvine, D.S., and Chatterjee, R. (2002) Sperm DNA damage in potentially fertile homozygous beta-thalassaemia patients with iron overload. *Hum Reprod*, 17, 1820–5.

Plowman, M.C., Grbac-Ivankovic, S., Martin, J., Hopfer, S.M., and Sunderman, F.W.J. (1994) Malformations persist after metamorphosis of Xenopus laevis tadpoles exposed to Ni2+, Co2+, or Cd2+ in FETAX assays. *Teratog Carcinog Mutagen*, 14, 135–44.

Plowman, M.C., Peracha, H., Hopfer, S.M., and Sunderman, F.W.J. (1991) Teratogenicity of cobalt chloride in Xenopus laevis, assayed by the FETAX procedure. *Teratog Carcinog Mutagen*, 11, 83–92.

Ponnapakkam, T.P., Sam, G.H., and Iszard, M.B. (2003) Histopathological changes in the testis of the Sprague Dawley rat following orally administered manganese. *Bull Environ Contam Toxicol*, 71, 1151–7.

Sanchez, D.J., Domingo, J.L., Llobet, J.M., and Keen, C.L. (1993) Maternal and developmental toxicity of manganese in the mouse. *Toxicol Lett*, 69, 45–52.

Saxena, D.K., Murthy, R.C., Lal, B., Srivastava, R.S., and Chandra, S.V. (1990) Effect of hexavalent chromium on testicular maturation in the rat. *Reprod Toxicol*, 4, 223–8.

Schardein, J. (1993). *Chemically Induced Birth Defects.* Marcel Dekker, New York.

Schroeder, H.A. and Mitchener, M. (1971) Toxic effects of trace elements on the reproduction of mice and rats. *Arch Environ Health*, 23, 102–6.

Smith, M.K., George, E.L., Stober, J.A., Feng, H.A., and Kimmel, G.L. (1993) Perinatal toxicity associated with nickel chloride exposure. *Environ Res*, 61, 200–11.

Sunderman, F.W., Jr., Shen, S.K., Mitchell, J.M., Allpass, P.R., and Damjanov, I. (1978) Embryotoxicity and fetal toxicity of nickel in rats. *Toxicol Appl Pharmacol* 43, 381–90.

Sylvester, S. and Griswold, M. (1994) The testicular iron shuttle: a "nurse" function of the Sertoli cells. *J Androl*, 15, 381–5.

Szakmary, E., Ungvary, G., Hudak, A., Tatrai, E., Naray, M., and Morvai, V. (2001) Effects of cobalt sulfate on prenatal development of mice, rats, and rabbits, and on early postnatal development of rats. *J Toxicol Environ Health A*, 62, 367–86.

Tarnacka, B., Rodo, M., Cichy, S., and Czlonkowska, A. (2000) Procreation ability in Wilson's disease. *Acta Neurol Scand*, 101, 395–8.

Treinen, K.A., Gray, T.J., and Blazak, W.F. (1995) Developmental toxicity of mangafodipir trisodium and manganese chloride in Sprague-Dawley rats. *Teratology*, 52, 109–15.

Trivedi, B., Saxena, D.K., Murthy, R.C., and Chandra, S.V. (1989) Embryotoxicity and fetotoxicity of orally administered hexavalent chromium in mice. *Reprod Toxicol*, 3, 275–8.

Webster, W.S. and Valois, A.A. (1987) Reproductive toxicology of manganese in rodents, including exposure during the postnatal period. *Neurotoxicology*, 8, 437–44.

Wei, Q., Fan, R., Yang, X., and Chen, T. (2003) Effect of zinc on reproductive toxicity in rats. *Wei Sheng Yan Jiu*, 32, 618–619.

Zahid, Z., Al-Hakkak, Z., Kadhim, A., Elias, E., and Al-Jumaily, I. (1990) Comparative effects of trivalent and hexavalent chromium on spermatogenesis of the mouse. *Toxicol Environ Chem*, 25, 131–6.

chapter 6

Lead Exposure and Its Effects on the Reproductive System

Rebecca Z. Sokol
Department of Obstetrics and Gynecology and Medicine
Keck School of Medicine, University of Southern California

Contents

Lead use dates back to antiquity. The Egyptians and the ancient Chinese emperors used lead in the production of ornaments and eye paints. Lead was used as a by-product of smelting silver in 2000 B.C., and the Romans used lead to make their aqueducts and water mains. They cooked and prepared their wine in lead-lined pots (Cunningham, 1986).

Greek physicians described lead poisoning over 2000 years ago. Both the Greeks and the Romans recognized lead as an abortifacient and a reproductive toxicant. Some historians attribute the fall of the Roman Empire to lead toxicity (Gillifilan, 1965; Cunningham, 1986).

Since the 1970s, animal and clinical studies have documented reproductive toxicity in both men and women exposed to lead in their workplace and in their environment.

Sources and Routes of Exposure

Although lead occurs naturally in the earth's crust, water, and air, lead in the atmosphere is primarily produced as a by-product of industry and gasoline. Lead sources are plentiful in industrialized nations. Until the past decade, the primary source of lead exposure was lead-based gasoline. Due to strict governmental controls in the United States, this exposure has dropped dramatically. Nonetheless, 890,000 children still have elevated levels (Pirkle et al., 1998). Lead-contaminated industrial emissions continue in countries other than the United States.

Leaded paint in older houses, lead in dust and soil, lead-based pottery, and lead in folk medicines and nonprescription drugs continue to be important sources of lead exposure. Recent reports of significant lead toxicity in individuals involved in pottery production in villages in Mexico have been published (Hernández-Serrato et al., 2003; Hibbirt, 1999). These studies reported an average estimated total daily lead intake of 4.0 milligrams in adults and 4.3 milligrams for children living in a village where pottery production was extensive (Hibbirt, 1999). Best predictors for high blood lead concentrations by linear multiregression analysis were occupation, gender, and the use of glass stoneware (Hernández-Serrato et al., 2003).

Calcium supplements are another potential source of lead ingestion. Four of seven natural products have measurable lead contents amounting to approximately 1 microgram per day per 800 ml per day of calcium. Four of 14 refined products also had similar lead content. No lead was detected in calcium acetate products. (Ross et al., 2000).

Absorption and Metabolism

Routes of exposure to lead include inhalation, oral, and dermal. Children tend to have a higher intestinal absorption rate than adults and thus may obtain higher blood lead levels with lower intakes. Lead exists in three kinetic pools in the body, the skeleton, the soft tissues, and the blood. The skeleton is the primary site of storage (Hernandez-Avila et al., 1998). The lead content of bone makes up approximately 70 to 90% of the total body load in children and adults, respectively, and increases with age (Leggette, 1993). Most of the lead in blood is located in the red blood cells. It is rapidly transferred from the plasma to the erythrocytes where it is bound to the cell membranes and hemoglobin. Soft tissue pools of lead are located in the liver, kidney, and the brain. The half-life of lead in the blood is 27 days, in the soft tissues approximately 30 days, and in the bone at least 20 years (Rabinowitz et al., 1973). These three kinetic pools are in constant equilibrium with the soft tissues and the bones serving as sources for mobilization of lead into the plasma. Mobilization of blood from bone in women may occur at times of rapid bone turnover such as pregnancy, lactation, and menopause (Silbergeld et al., 1988; Silbergeld, 1991; Vahter, et al., 2004; Symanski and Hertz-Picciotto, 1995).

Physiology of Reproduction

The reproductive axis consists of the hypothalamus, pituitary, and gonad. The hypothalamic-pituitary-gonadal (HPG) axis is physiologically a closely regulated system. Control and coordination of reproductive function occurs via feedback signals, both positive and negative exerted by the hormones secreted at each level of the HPG axis. These signals include (1) stimulation of gonadotropin releasing hormone (GnRH) by neurotransmitters, (2) stimulation of luteinizing hormone (LH) and follicle stimulating hormone (FSH) by the pulsatile release of GnRH, (3) stimulation of the gonad by the gonadotropins, (4) inhibition of the hypothalamus and pituitary responsiveness to GnRH by gonadal steroids, and (5) inhibition of pituitary FSH release by inhibin secretion (Sokol, 1997).

Any disruption by lead of the delicately coordinated interactions among the components of the reproductive axis may lead to reproductive abnormalities.

The Male Reproductive Axis (Figure 6.1)

In man, the reproductive hormones ultimately regulate spermatogenesis. Human spermatogenesis takes place in the seminiferous tubules of the testes and covers a 72-day period (Clermont, 1986). Spermatozoa spend an additional 10 days moving from the rete testes into the epididymis. Following ejaculation and prior to achieving the ability to fertilize an egg, spermatozoa must undergo capacitation and the acrosome reaction, processes that result in the ability of the spermatozoa to enter the ova (Sokol, 1997).

LH plays an important role in initiation and maintenance of sperm production, primarily by stimulating testosterone secretion by direct action on the interstitial cells (Barraclough, 1982). FSH is also necessary for the initiation of spermatogenesis at the time of puberty (Plant, 1986). Inhibin is also involved in the feedback regulation of FSH secretion. Testicular testosterone exerts negative feedback on the HP (hypothalamic- pituitary) unit, with the administration of testosterone suppressing gonadotropin release and orchiectomy or testicular failure leading to an increase in the level of natural secretion (Matsumoto and Bremmer, 1987).

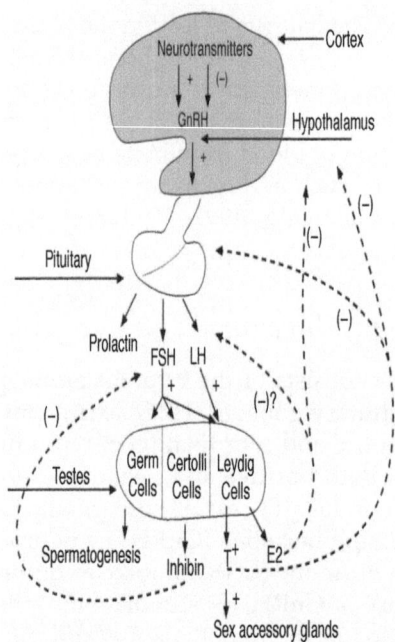

Figure 6.1 Hypothalamic-Pituitary-Testicular Axis. Sokol. Mishell's Textbook of Infertility, Contraception, and Reproductive Endocrinology: 548, 1997. Blackwell Science, MA.

The Female Reproductive Axis (Figure 6.2)

In the human female, the reproductive hormones maintain the menstrual cycle (Mishell et al., 1971). LH acts on the ovary to produce estradiol, testosterone, and progesterone. FSH stimulates follicular maturation. Estrogens and progesterone exert both positive and negative feedback control of GnRH and gonadotropin secretion. Ovulation depends on the ovary, which must maintain appropriate estradiol levels during the menstrual cycle (Shoupe and Lobo, 1997).

The menstrual cycle can be divided into three distinct phases: the follicular phase, the ovulatory phase, and the luteal phase. The follicular phase is characterized by the selection of the dominant follicle and growth of the endometrium under the influence of estrogen. The ovulatory phase is characterized by a switch from estrogen to progesterone production and the final release of an oocyte. The luteal phase is characterized by the preparation of the endometrium for implantation of the corpus luteum. LH has multiple target sites, whereas FSH appears to act only on the granulosal cells. The dominant follicle, as it develops and produces increasing amounts of estrogen,

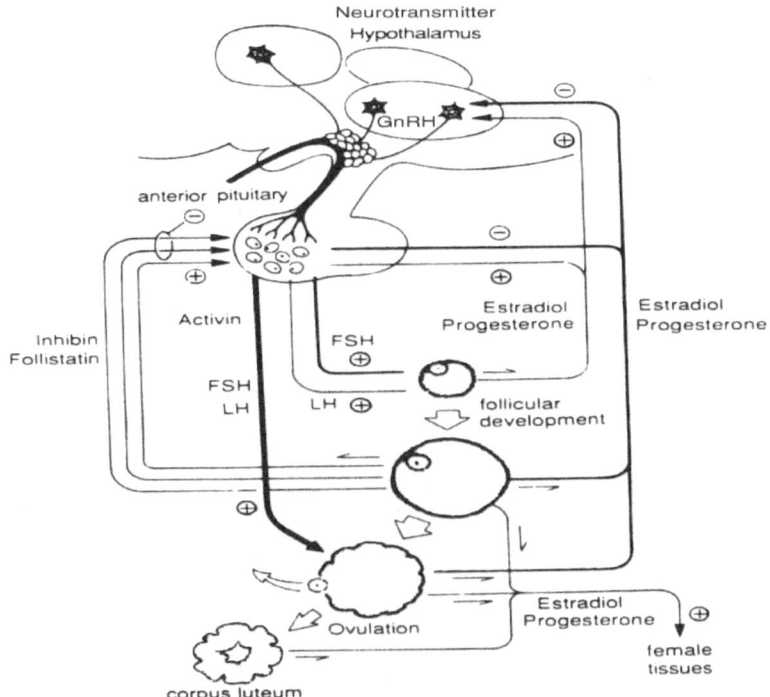

Figure 6.2 Hypothalamus-Pituitary-Ovarian Axis. Modified from Lobo et al. Mishell's Textbook of Infertility, Contraception, and Reproductive Endocrinology: 125, 1997. Blackwell Science, MA.

exerts a positive feedback on the hypothalamic-pituitary axis (Hoff et al., 1983) and the endometrium response to ovarian steroid production by increasing the glandular epithelium and endometrial blood vessels. Absence of implantation and subsequent production of HCG results in corpus luteum involution, and estrogen and progesterone levels decline.

Effects of Lead on Male Reproduction

Although the vast majority of the early published literature on the effects of lead on reproduction primarily addressed adverse effects on female reproduction, the majority of studies over the past 3 decades have evaluated adverse effects on male reproduction (Rom, 1976; Bell and Thomas, 1980; Winder, 1989; Sokol, 1997; Apostoli et al., 1998; Sokol, 1998; Health Council, 2003). In one of the first published studies, dating to the 1850s, De Quatrefages reported that heavy metals, including lead, interfered with sperm motility (Mann, 1981). More recently, lead has been implicated as toxic at every level of the male reproductive system. However, a few studies question if lead is indeed a direct reproductive toxicant. The majority of studies have been conducted using the animal model. Clinical studies appear to substantiate the findings of the animal studies.

Animal Studies

Animal studies have been conducted in mice, rats, rabbits, and monkeys with variable results (Table 6.1). These variable results have been attributed to differences in study design and strain differences within and between species in susceptibility to lead toxicity (Apostoli et al., 1998). Nonetheless, the majority of animal studies suggest that lead exposure disrupts the HPG axis in males.

In 1973, Hildebrand and colleagues reported that male adult rats dosed with lead acetate in their drinking water for 30 days had evidence of atrophy of the seminiferous tubules, with inhibition of spermatogenesis and decreased sperm motility (Hildebrand et al., 1973). Additional studies on rats and mice, with some exceptions, confirmed alterations in spermatogenesis and sperm motility at lead doses attaining blood lead concentrations of greater than 30 to 40 µg/dl with dosing durations of greater than 14 days (Table 6.1).

A number of studies have evaluated the impact of lead exposure on the reproductive hormones. Most studies noted a suppression of testosterone levels, circulating and/or intratesticular (Sokol et al., 1985; Rodamilans et al., 1988; Nathan et al., 1992; Thoreaux-Manley et al., 1995; Ronis et al., 1996; Moorman, 1995). FSH and/or LH levels and content were reported to be increased (Petrusz et al., 1978); unchanged (Sokol et al., 1985; Nathan et al., 1992; Foster et al., 1993; Ronis et al., 1996), and suppressed (Wiebe, 1983; Sourgens et al., 1987; Thoreux-Manley et al., 1995) in response to lead dosing. Some studies have reported a direct spermatoxicity without hormonal disruption, whereas others report no effect of lead on reproductive hormones or spermatogenesis (Table 6.1).

In a series of experiments designed to determine the pathophysiology of lead's toxic effects on the male reproductive axis, male rats dosed with lead-containing water were studied in a variety of experimental paradigms. In an initial dose-response study, Wistar rats were sacrificed after 30 days of exposure. Significant negative correlations between whole blood lead level and sperm concentration and serum and intratesticular testosterone values were found. FSH levels were significantly suppressed, but LH levels were not, suggesting that the primary site of toxicity was at the hypothalamic pituitary level of the HPG axis (Sokol et al., 1985). To identify more specifically the site of lead's toxic actions on the HPG axis, the response of lead-dosed animals to hormone stimulation was studied (Sokol, 1987). To determine if the primary site of toxicity was the testicle, the animals were

Table 6.1

Reference[a]	Species Strain Age at Start	Lead Compound	Route of Administration	Duration of Exposure Dosage	Blood Pb-Level at Sacrifice[b] (µg/dL)	Results
Al-Hakkak (1988)	Mouse Balb-C Albino Swiss Weanling	Lead monoxide alloy powder employed in Iraq	Oral Feed	35, 70 days Dosage: 25 and 50 mg/kg	N/A	Control females mated with lead-ingested males showed a decrease in implantation. Mean nonviable pregnancies in the exposed groups were sig. higher than control. Exposure increased the frequency of lethal mutations in the sperm, and post-implantation loss. Number of cells at zygotene stage increased, but at all other substages, showed a consistent decrease, particularly spermatagonia and spermatocytes.
Barratt (1989)	Rat Wistar 70 days	Lead Acetate	Drinking water and gavage	63 days Dosage: 0–330 mg/kg/d	N/A	Decreased abnormal spermatogenesis.
Batra (1998)	Rat N/A N/A	Lead Acetate	Orally	3 months	N/A	Zinc supplementation ameliorates lead-induced testicular damage at cellular and sub-cellular level.
Boscolo (1988)	Rat Sprague-Dawley Weanling	Lead Acetate	Drinking water	18 months Dosage: 60 ppm	3.9–16.7	Lead was increased in blood, kidneys, and brain, but not in testis. Sertoli cells increased size of lysosomes, all else was normal.
Chowdhury (1984)	Rat Albino Mature	Lead Acetate	Drinking water	60 days Dosage: 0.0, 0.25, 0.5, 1.0 g/l	8–142	Testicular atrophy and cellular degeneration. Partial inhibition of spermatogenesis.
Chowdhury (1986)	Rat Charles River Mature	Lead Acetate	Intraperitoneal injections	30 days Dosage: 0, 1, 2, 4, 6 mg/kg	5–35	Increase in blood and testis lead levels, decrease of ALA-D levels. Changes in testicular tissue and abnormal spermatogenesis.

Table 6.1 (continued)

Reference[a]	Species Strain Age at Start	Lead Compound	Route of Administration	Duration of Exposure Dosage	Blood Pb-Level at Sacrifice[b] (µg/dL)	Results
Chowdhuri (2001)	Mouse N/A N/A	Lead Acetate	N/A	2 hours	N/A	Low frequency of sperm attachment; leads to reproductive failure.
Coffigny (1994)	Rat Sprague-Dawley N/A	Lead Oxide *In utero* exposure	Inhaled	13 days during gestation. Dosage: 5 mg/m^3	71.1–83.2	No effect on hormones, sperm fertility at 90 days of age.
Corpas (1998)	Rat Wistar	Lead + Cadmium	Drinking water	During gestation and early lactation, until 5 days after parturition. Dosage: 10 mg/L	N/A	Reduced DNA gonadal content, reduction in number of spermatagonia.
Der (1976)	Rat Sprague-Dawley N/A	Lead Acetate	Intraperitoneal injections	70 days. Dosage: 0, 50, 250	23–73	No histological change in testes or seminal vesicles.
Easwari (1992)	Rat N/A N/A	Lead Acetate	Drinking water	10 weeks	2.3	No effect.
Eyden (1978)	Mouse Balb/C 90 days	Lead Acetate	Drinking water and food	8 weeks	N/A	Infertility.
Foote (1999)	Rabbit N/A N/A	Lead	N/A	N/A	N/A	Reduced fertilization rate.

Study	Species	Compound	Route	Duration	Dosage	Days	Effects
Foster (1993)	Monkey Cynomolgus At birth	Lead Acetate	Drinking water	9 yrs	Dosage: 1500 µg/kg/day	3–26 Beginning at birth, postnatal day 300; or days 0–400	Decreased T, no change LH and FSH; decrease in INH/FSH; blunted GnRH stimulation.
Fowler (1980)	Rat CD F_1 21 days F_2 Intrauterine	Lead Acetate	Gelatin capsule	6 or 9 months F_1 6–7 weeks	Dosage: 5–250 ppm	4–97	No sig. decrease in sperm count/morphology.
Hilderbrand (1973)	Rat Sessco Mature	Lead Acetate	Drinking water	30 days	Dosage: 0, 0.5, 10 µg/l	14–30	Increase of prostate twofold, seminal vesicle and testes size normal. Seminiferous tubule damage, spermatogenesis halted.
Hsu (1998)	Rat N/A N/A	Lead Acetate	Intraperitoneal injections	6 weeks		N/A	Decreased serum testosterone level; early onset of capacitation.
Johansson (1986)	Mouse NMRI Mature	Lead Chloride	Drinking water	12 weeks	Dosage: 0, 1 g/l	0–32	No difference in blood T and epididymal sperm count. Decreased number of pregnant females.
Johansson (1987)	Mouse NMRI N/A	Lead Chloride	Drinking water		Dosage: 1 g/L	N/A	Delayed hatching, lower hatching ability, decrease in inner cell mass development seen in mice fertilized by lead-exposed mice.
Johansson (1989)	Mouse NMRI 63 days	Lead Chloride	Drinking water	16 weeks	Dosage: 1g/ L	0.05–40	Increased frequency in acrosome-reacted spermatozoa. Number of spermatozoa bound to zona pellucida of denuded mouse eggs increased. Ability of spermatozoa of exposed mice to penetrate zp oocyte sig. reduced. Motility not affected.
Kempinas (1994)	Rat Wistar 50 days	Lead Acetate	Drinking water and i.v.	20 days	Dosage: 1.0 g/l and 0.1 µg/100 L	10–41	Variable effects.

Table 6.1 (continued)

Reference[a]	Species Strain Age at Start	Lead Compound	Route of Administration	Duration of Exposure Dosage	Blood Pb-Level at Sacrifice[b] (µg/dL)	Results
Klages (1987)	Rat Wistar N/A	Lead Acetate	Drinking Water	30 days Dosage: 0.0045%, 0.09%, 0.45%	5.0–60	No lead in testes. Increased lead content in brain, FSH and T no change, decrease in prolactin and LH.
Klein (1994)	Rat Sprague-Dawley 100 days	Lead Acetate	Drinking water	21 days Dosage: 0.1, 0.3, 0.6%	5–55	Suppression of HPT axis in male rats. Positive correlation between BLL[c] and increase in intracellular levels of mRNA and stored hormones.
Krasovskii (1979)	Rat White N/A	Lead Acetate	Drinking Water	20–30 days or 6–12 months Dosage: 0.0015–0.05 mg/kg	N/A	Sperm motility decreased.
McGivern (1991)	Rat Sprague-Dawley *In utero*—mature	Lead Acetate	Drinking water	Day 14 of gestation to parturition Studied and/or sacrificed at 70, 100, 160 days Dosage: 0.1%	In weaned pups (not detectable)	Male offspring; decreased sperm count (70 and 160 days of age). Azoospermia observed in one lead-exposed animal (70 days), two animals (160 days). Enlarged prostates in males (160 days), other sex wt. organs normal. Volume of sexually dimorphic nucleus of the preoptic area of hypothalamus in adulthood sig. reduced. Pulsatile release of gonadotropins, measured in castrated animals (both sexes) revealed irregular release patterns of FSH and LH. Abnormal reproductive behavior in adult males.
Nathan (1992)	Rat Sprague-Dawley 70 days	Lead Acetate	Drinking water	10 weeks Dosage: 0, 0.05%, 0.1%, 0.5%, 1.0 %	2.3–124	Serum, FSH, LH, T, and inhibition levels normal, as well as pituitary content of LH and FSH. No disruption of spermatogenesis. Increase in epididymal androgen-binding protein level, decrease in seminal vesicle weight observed 1% ingested rats.

Reference	Species/Strain	Compound	Route	Duration	Dosage	Blood Lead	Effects
Petrusz (1979)	Rat Long Evans Neonatal	Lead Acetate	Gastric gavage	10, 15, 20 days		7–1055	Increased pituitary FSH content, no change in serum FSH. No change pituitary LH content.
Piasek (1987)	Rat Albino N/A	Lead Acetate	Drinking water	18 weeks	Dosage: 0–200 mg/kg/day	N/A	No effect on fertility in first or second mating. No paternal reproductive effects observed.
Pinon-Lataillade (1993)	Rat Sprague-Dawley N/A	Lead Acetate	Drinking water/ inhalation	70 days	Dosage: 1500, 3500, 5500 ppm	51.1–58	Lead acetate ingestion did not affect reproductive system or fertility. Inhalation did not affect fertility either, but seminal vesicle weight dropped.
Pinon-Lataillade (1995)	Mouse NMRI From weanling	Lead Acetate	Drinking water	60 days	Dosage: 0.3% or Lead Oxide 5 mg m^{-3}; Dosage: 0.5%	4–132	When mated with unexposed mice, reproductive hormones, spermatogenesis, and male fertility not affected. Lead-exposed males had smaller weight testes by 13%, seminal vesicle and prostate weights were diminished by 29%.
Rodamilans (1988)	Mouse BALB/C 63 days	Lead Acetate	Drinking water	6 months	Dosage: 366 mg/l	10–67	Lead conc. in testes increased significantly on day 60. Reduction in intratesticular testosterone levels after 30 days exposure and of androstenedione levels after 150 days.
Ronis (1994)	Rat Sprague-Dawley Newborn	Lead Acetate	Drinking water	85 days		0–316	Decreased male secondary sex organ weight. 70% suppression of circulating testosterone in male, estadiol suppression to undetectable levels, decreases in LH. Reproductive toxicity accompanied by 25% decrease in prepubertal growth rates from 2.6 to 1.8 g/day. During puberty only males exhibited delayed growth rates.

Table 6.1 (continued)

Reference[a]	Species Strain Age at Start	Lead Compound	Route of Administration	Duration of Exposure Dosage	Blood Pb-Level at Sacrifice[b] (µg/dL)	Results
Ronis (1996)	Rat Sprague-Dawley Newborn Prepubertal Postpubertal	Lead Acetate	Drinking water	85 days (in utero, prepubertally, postpubertally)	0–316	*In utero*: T suppressed, severe reproduction disruption, prepubertal growth suppressed 25%. Prepuberty: secondary sex organ weights decreased. Postpuberty: variable effects on circulating T, pit. LH and pit LH beta mRNA.
Ronis (1996)	Rat Sprague-Dawley Newborn	Lead Acetate	Drinking water	Dosage: 0, 0.6% 21, 35, 55, and 85 days Dosage: 0%, 0.05%, 0.15%, and 0.45%	0–214	Prostate weights reduced at age 35, but caught up to controls by age 85. Reduced circulating sex steroid levels at birth, decrease in T and estradiol, during puberty. By age 85, sex steroid levels and male epididymal sperm counts did not differ from controls even at highest lead dosage. Variable response to GnRH challenge.
Ronis (1998)	Rat Sprague-Dawley Gestation	Lead Acetate	Drinking water	During gestation and through day 21, 35, 55, 85	0–379	Birth weight, prepubertal and pubertal growth rates sig. suppressed. Decrease in sex steroids (T) and estradiol observed during puberty. Postpuberty, sex steroid levels back at control levels. No effect on sperm count.
Saxena (1984)	Rat ITRC N/A	Lead Acetate	Intraperitoneally	15 days Dosage: 8mg/kg	N/A	Degeneration of seminiferous tubules and hypospermatogenesis.
Saxena (1986)	Rat ITRC N/A	Lead Acetate	Intraperitoneally	15 days Dosage: 0, 5, 8, 12 mg/kg/day	0.071–15.7	Sig. increase in testis size shown in group exposed to 8 and 12 mg. Total cholesterol level decreased for all exposed groups, no pathological changes observed in testis.
Saxena (1987)	Rat Wistar 21 days	Lead Acetate	Intraperitoneally	120 days Dosage: 0, 8mg/kg	N/A	Abnormal spermatogenesis, Leydig cell degeneration. Possible disturbance of testicular spermatogenesis.

Study	Species	Compound	Route	Duration	Dosage		Results
Sokol (1985)	Rat Wistar 52 days	Lead Acetate	Drinking water	30 days	Dosage: 0, 0.1, 0.3 %	4–60	Negative correlation with T and BLL, sperm counts sig. decreased, LH no sig. diff., sig. increase in ventral prostate weight, FSH suppressed.
Sokol (1987)	Rat Wistar 52 days	Lead Acetate	Drinking water	30 days	Dosage: 0%, 0.3%	5–30	Hyperresponsiveness to GnRH stimulation, blunted response to naloxone.
Sokol (1989)	Rat Wistar 27, 52 days	Lead Acetate	Drinking water	30 days	Dosage: 0 or 0.6 %	5–58	At end of 30-day recovery period, serum testosterone and serum parameters normalized in prepubertal animals, but not in pubertal animals.
Sokol (1990)	Rat Wistar 52 days	Lead Acetate	Drinking water	7, 14, 30, or 60 days	Dosage: 0 or 0.6 %	8–75	Lead and FEP levels sig. higher than controls, testosterone and spermatogenesis suppressed in lead-exposed group (except those treated for 7 days). Increased duration of exposure after 14 days does not further suppress T levels or spermatogenesis.
Sokol (1991)	Rat Wistar 42, 52, 70 days	Lead Acetate	Drinking water	30 days	Dosage: 0.1 or 0.3%	7–60	T and sperm conc. and production rates suppressed (most sig. in 52 and 70 days), prepubertal animals least vulnerable.
Sokol (1994)	Rat Wistar 100 days	Lead Acetate	Drinking water	14, 30, 60 days	Dosage: 0 or 0.3%	5–46	Lead exposure *in vivo* disrupts sperm's ability to penetrate/fertilize eggs collected from nonexposed females *in vitro*.
Sokol (1998)	Rat Sprague-Dawley 100 days	Lead Acetate	Drinking water	7 days	Dosage: 0 or 0.3%	8–74	AMPT dose had no sig. effect on LH secretion in controls, but LH fell sig. in lead-treated animals. High dose rate had suppressed LH and FSH pulsatility.
Sokol (2002)	Rat Sprague-Dawley 100 days	Lead Acetate	Water	1, 4, 8, 16 weeks	Dosage: 0, 0.025, 0.05, 0.1, 0.3%	2–42	No sig. diff. in LH and GnRH in treated animals. 1-week treatment: dose-related increase in GnRH mRNA conc.

Table 6.1 (continued)

Reference[a]	Species Strain Age at Start	Lead Compound	Route of Administration	Duration of Exposure Dosage	Blood Pb-Level at Sacrifice[b] (µg/dL)	Results
Sourgens (1987)	Rat Wistar N/A	Lead Acetate	Drinking water	30 days Dosage: 0.0%, 0.0045%, 0.09%, 0.45%	5–60	PRL and LH depressed. T and E2 unaltered. Increased testicular lead in high-dosed gonads, no histological changes.
Thoreaux-Manlay (1995)	Rat Sprague-Dawley 97 days	Lead Acetate	Intra-peritoneal injection	5 weeks / 5 daily treatments	4–1700	No sig. effects on spermatozoa production, motility, and morphology. Prolactin reduced 30%, LH reduced 32%, T reduced 77%. Did not affect fertility of males.
Varma (1974)	Mouse N/A N/A	Lead Acetate	Drinking water	4 weeks	N/A	Spermatotoxicity, infertility.
Wadi (1999)	Mouse CF-1 49 days	Lead Acetate	Drinking water	6 weeks Dosage: 0.2% Dosage: 0.25% and 0.5%	N/A	FSH, LH, and testosterone not affected; decreased epididymus and seminal vesicle, reduced number of sperm within epididymis (low-lead dose) and high-lead dose reduced sperm count and percentage of motile sperm.
Wiebe (1982)	Rat Sprague-Dawley In utero	Lead Acetate	Drinking water	Pregnant dams prenatally exposed through 13 or 21 days	3.4–6.3	Steroid production and hormone binding in testis reduced at time of puberty.
Wiebe (1983)	Rat Sprague-Dawley Mature	Lead Acetate	Sertoli cells cultured *in vitro*	1, 4, 24, 48, 96, 144 hr dose response	N/A	After 24 hr exposure, cells exhibited 10–20% decrease in FSH binding, and cyclic AMP production; after 96 hr 75% decrease in two parameters; no reduction seen in 1–4 hr period. Inhibition greater in younger (16-day-old) rats than in 20-day-old rats. 25% decrease in testicular enzymes.

Willems (1982)	Rabbits New Zealand Mature	Lead Acetate	Injection	14 weeks Dosage: 0, 0.25, 0.5 mg/kg	0.32–2.97 umol/L	No effects on testis histology, sperm count, or morphology.
Zirkin (1985)	Rat Long Evans Mature	Lead Acetate	Drinking water	130 days Dosage: 0–8000 ppm	N/A	Decreased testosterone, no effect on spermatogenesis.

[a] First author

[b] Unless otherwise indicated

[c] Blood lead levels

stimulated with human chorionic gonadotropin (hCG), a hormone that stimulates testosterone production in the testis in a manner identical to the endogenous hormone, LH. Lead-treated animals responded to hCG stimulation in a hyperresponsive manner, indicating an intact pituitary-testicular axis. To determine if lead's toxicity was pituitary directed, the animals were stimulated with GnRH. The lead-treated animals released more LH in response to the GnRH than did the control animals, suggesting that the hypothalamic-pituitary axis was intact. The enhanced response to both GnRH and LH indicated that hormone production was intact, but that release of the hormone was attenuated in a lead-exposed animal. At the molecular level, a significant dose-dependent increase in mRNA levels of GnRH and B-LH, and a significant increase in pituitary stores of B–LH occurs in lead-dosed rats (Klein et al., 1994). Increased release of LH from pituitaries harvested from lead-treated animals produce more GnRH-stimulated LH under cell culture conditions than do pituitaries harvested from control animals (Sokol et al., 1998). Unlike the rat studies, a blunted response to GnRH was reported following long-term lead exposure in male monkeys, as was a decrease in the inhibin/FSH ratio (Foster, 1993).

Studies conducted on rabbits suggest dual sites of toxicity, both at the hypothalamic-pituitary levels and in the testicle. An inverse relationship between blood lead and sperm count and testosterone was reported in mature male rabbits treated with subcutaneous injections of lead acetate for 15 weeks. Unlike the rat studies discussed, FSH levels were marginally increased, but the pattern of LH response was variable. Abnormalities of sperm morphology were noted on light microscopy, but not EM (electron microscopy) (Moorman et al., 1995). Similarly, no ultrastructural changes on EM of spermatozoa were noted in lead-treated rats, but lead treated groups fertilized fewer ova *in vitro* (Sokol et al., 1994). Lead treatment of sexually mature male CF-1 mice for 6 months did not alter reproductive hormones, but did decrease epididymal sperm numbers and percentage of abnormal sperm (Wadi and Ahmed, 1999).

Lead exposure does disrupt sperm function and may decrease fertility. Evidence supporting a disruption in fertility *in vivo* is mixed. Early studies reported lead exposure–related infertility in rats and mice (Puhac et al., 1963; Varma et al., 1974). In contrast, lead-treated male Sprague-Dawley rats that were found to have reduced plasma concentrations of LH and testosterone did not have reduced fertility as measured by the number of pregnant females, as compared to control animals (Thoreux-Manlay et al., 1995). Similar results were reported by Zirkin and coworkers (Zirkin et al., 1985). Similarly, epididymal spermatozoa concentration, morphology, and fertility were not adversely affected in lead-exposed male mice offspring of lead-exposed dams mated to unexposed females (Pinon-Lataillade and Thoreaux-Manley, 1995).

Rodents produce large amounts of spermatozoa, which may reduce the ability of multigeneration studies to uncover fecundity changes. A chemically induced reduction in daily sperm production is unlikely to alter fecundity unless sperm counts are suppressed by 80% (Bechter et al., 1982). Alternative methods of studying the impact of lead exposure on fertility include *in vitro* fertilization conditions and artificial insemination protocols. A number of studies employing these approaches support the hypothesis that lead exposure alters fertility.

Using a mouse model, Johansson and coworkers reported that long-term exposure to inorganic lead reduced the ability of sperm harvested from lead-exposed male mice to fertilize mouse eggs *in vitro* (Johansson et al., 1987). An increased frequency of acrosome-reacted spermatozoa, delayed hatching, a lower hatching ability, and a decrease in the inner cell mass development of blastocysts was also reported (Johansson et al., 1987; Johansson, 1989). Studies using the rat model reported similar findings. Sperm harvested from lead-treated animals penetrated fewer eggs in an *in vitro* system, and fewer of the fertilized eggs progressed through the stages of fertilization. No ultrastructural changes were noted in the spermatozoa harvested from the lead-treated animals, suggesting a functional defect (Sokol et al., 1994).

Experiments conducted in rabbits agree with the rodent studies. Fertilization rates were reported to be significantly lower in does inseminated with lead exposed spermatozoa (Foote, 1999). Male rabbits exposed to lead *in utero* produced spermatozoa in adulthood that manifested acrosomal dysgenesis and nuclear malformations (Veeramachaneni et al., 2001).

Recently, studies have focused on the underlying mechanisms accounting for the disruption of sperm–egg interaction associated with lead exposure, particularly at the molecular level. Chowdhuri and colleagues (Chowdhuri et al., 2001) utilized a mouse *in vitro* system to ascertain the role that lead may play on gamete physiology at the molecular level during the sperm–zona binding process. They found that DNA, RNA, and protein synthesis under sperm-binding conditions were affected following incubation of ova and sperm in a lead-rich culture medium. Hsu and coworkers (1998) reported an association between lead exposure, excess production of sperm-reactive oxygen species, and premature acrosome reaction and reduced zona-intact oocyte-penetrating capability. Preliminary evidence also implicates a decrease in testicular RCK4 expression, which is involved in the regulation of acrosome function, and which may play a role in lead-induced decreased sperm fertilizing potential (Millan et al., 2003). Lead was also noted to decrease the level of HP2-DNA interaction, probably by competing with or replacing the zinc ion (Benoff et al., 2003). Lead-induced alterations in calcium homeostasis at the testicular cell level may also play a role in lead-induced spermatotoxicity (Benoff et al., 2000).

Age, maturity, animal strain, and the ability of the animal to resist the toxic effects of lead exposure may be reasons why some studies report clear-cut toxicity and others do not. Prepubertal rats are less sensitive to the toxic effects of lead than animals with exposure to lead beginning after puberty (Sokol and Berman, 1991; Ronis et al., 1996, 1998). This is particularly true for animals exposed *in utero* and postpartum. In male pups, significantly reduced serum luteinizing hormone (LH) and testosterone were observed at puberty. However, by age 85 d, plasma sex steroid levels and male epididymal sperm counts did not differ from controls even at the highest lead dose (Ronis et al., 1996, 1998).

Sprague-Dawley rats are more resistant than Wistar or Sesco rats, and certain strains of mice might be more vulnerable (Schroeder and Mitchner, 1971; Der et al., 1976; Apostoli et al., 1998; Benoff et al., 2003). Toxicity seems to be reversible, and animals appear to adapt to the toxicity of lead. Hormone levels and spermatogenesis return to baseline levels after cessation of exposure to lead-containing water (Sokol, 1989). Prolonged exposure to lead does not further suppress serum testosterone, spermatogenesis or GnRH mRNA in the rat and monkey (Sokol, 1990; Ronis et al., 1997; Sokol et al., 2002; Foster, 1993).

Collectively, these findings support the hypothesis that lead interferes with the normal release of tropic hormones and disrupts hormone feedback mechanisms, and are consistent with the earlier studies reporting suppressed testosterone and/or spermatogenesis in combination with normal or low levels of gonadotropins. The primary site of toxicity appears to be in the CNS, although direct testicular toxicity may also occur. Pubertal rats are the most sensitive. Toxicity appears to be reversible and the animal adapts to toxicity with increasing duration of exposure (Sokol, 1998).

Clinical Studies

The first comprehensive study of the effects of lead exposure on the human male reproductive axis was reported by Lancranjan and coworkers (Lancranjan et al., 1975). Semen quality, serum testosterone levels, and urinary gonadotropins were measured in men who worked in a battery plant in Romania and were exposed to lead on the production line. The results were compared to those collected from men who worked in a low-lead office environment. The 150 men studied were divided into four exposure groups. A dose–response relationship was identified between blood lead level and a decrease in sperm concentration, sperm motility, and percentage of normal sperm morphology. Lower testosterone levels were measured in the men with the higher blood lead levels. There was no change in urinary gonadotropin levels, leading the authors to conclude that lead was primarily a testicular toxicant. However, a finding of inappropriately low testosterone levels in combination with low or normal gonadotropin levels is more consistent with a hypothalamic-pituitary directed toxicant.

During the almost 3 decades since the publication of the Lancranjan paper, a number of studies have investigated the effects of lead exposure in the clinical setting The majority of the studies reported a significant association between increasing lead levels in lead-exposed men and decreased sperm concentrations. The impact of lead exposure on the reproductive hormonal axis is less consistent (Table 6.2).

Table 6.2

Reference[a]	Subjects	Route of Administration	Duration of Exposure Dosage	Blood Pb-level Measured[b] (µg/dl)	Results
Alexander (1996)	2469 men surveyed 929 responded 152 serum 119 SA	Employees at Pb smelter	15.7–19.8 yrs Dosage: <15 to >40 µg/dL	5–58	Sperm count, conc., motile conc. inversely related to BLL_3[c].
Al-Hakkak (1986)	Control (N = 19) Exposed (N = 19)	Worked in a battery plant	4–14 yrs	Controls 24 Exposed 64	Rate of spontaneous abortion higher in wives of exposed men than controls. Increase in chromosome abnormalities in exposed men.
Assenato (1987)	Control (N = 18) Exposed (N = 18)	Worked in a storage battery plant	1–10 years Dosage: 0.054– 0.584 mg Pb/m3	Controls 18 Exposed 61	38% lower median sperm count in exposed. No significant difference in LH, FSH, PRL, T.
Benoff (2003)	Male human semen donors of proven fertility (N = 9) and partners of IVF patients (N = 140)	Unknown	N/A	N/A	Direct negative correlation between seminal plasma lead levels and IVF rates, lead levels not related to circulating levels of FSH, LH, and T. Lead levels positively correlated with spontaneous acrosome reaction and negatively correlated with mannose receptors.
Braunstein (1978)	Age 30–39 N = 10	Exposed and poisoned men working in lead smelter	1–3 years	29–88	2 men with oligospermia: no significant diff. in sperm motility or % abnormal forms. After administering human chorionic gonadotropin, increase of T was significantly greater in Pb-poisoned men and LH decreased, increase of estradiol decreased in Pb exposed.

Table 6.2 (continued)

Reference,[a]	Subjects	Route of Administration	Duration of Exposure Dosage	Blood Pb-level Measured[b] (µg/dl)	Results
Chowdhury (1986)	Control (N = 10) Exposed (N = 10)	Exposed working in a newspaper printing press	10 years Dosage: 8h/day	42.5	Less motile sperm, tail abnormalities increased, average sperm count lower.
Coste (1991)	Control (N = 125) Exposed (N = 229)	Worked in battery factory	5 years Dosage: Group 2: BLL <40, Group 3: BLL 60 Group 4: BLL > 60	Controls 8.3 Exposed 35.2	Lead exposure not associated with infertility. LH, FSH increased, T lower than controls.
Cullen (1984)	N = 7	Occupational exposure	5 weeks–15 years	66–139	6 patients had normal serum T conc., 5 patients had defects in spermatogenesis.
De Rosa (2003)	N = 85/ grp Case control study	Men employed at motorway tollgates	6.2 +/- 1.6 years	Controls: 7.4 +/- 0.5 Exposed: 20.1 +/- 0.6	Sperm count, LH and FSH, T all within normal ranges. Total motility, forward progression, functional tests, and sperm kinetics lower. Lead level inversely correlated with sperm count.
Fischer-Fischbein (1987)	Case report of one lead-poisoned man	Exposure as firearms instructor	6 years	88; after chelating therapy dropped to 30	After 2.5 years of therapy, sperm density and total sperm count increased, motility remained equal, head defects decreased, conceived a child after 1 year therapy.
Gennart (1992a)	Control (N = 138) Exposed (N = 74)	Worked in lead acid battery factory	1–28 years	Control: 4–19 Exposed: 24–75	Decrease in fertility.
Gennart (1992b)	Control (N =85) Exposed (N= 98)	Worked in lead acid battery factory	1–28 years	14–75	No effect on LH, FSH, TFT.
Govoni (1987)	N = 76	Worked in pewter factory	N/A	14–80	Increase in PRL.

Study	N	Exposure	Duration	Blood lead level	Results
Gustafson (1989)	Controls (N=25) Exposed (N = 25)	Worked at a secondary Pb smelter	10 mo–6.5 years	Controls: 5 Exposed 49 +/– 5	Decrease in FSH, LH.
Hu (1992)	Control (N = 24) Exposed (N = 24)	Worked in small printing house/ battery factory	N/A	Urine lead 87.6 µg/l	Sperm count decreased, teratospermia increased.
Lancranjan (1975)	Control (N = 50) Exposed (N = 100)	Working in a storage battery plant	N/A	23–75	BLL_3 inversely associated with decreased sperm count, teratospermia. Decreased T. No change in urinary gonadotropins.
Lerda (1992)	Control (N = 30) Exposed (N = 38)	Exposed in battery factory	Mean of 11.7 years	18–86.6	Sperm count and motility decreased and % of anomalies increased with increasing lead level.
Lin (1996)	Control (N = 5148) Exposed (N = 4256)	New York state heavy metals registry	N/A	25.0–50.0	Exposed workers had fewer births, not assoc. with fertility, men with the highest exposure (>50) or >5 yrs, significantly less likely to have any children.
Lindbom (1991)	Males with wives having spont. abortions (N = 213) or gave normal birth (N = 300)	Occupationally exposed	N/A	<21–39	Increased risk for spontaneous abortion found when BLL_3 > 31µg/ dl.
McGregor (1990)	N = 90	Occupational exposure to inorganic lead	1–45 years	11–77	Subclinical increase in FSH. No change in T, possible decrease in LH.
Ng (1991)	Control (N = 49) Exposed (N = 122)	Exposed men worked in Pb battery factories	0.1–19 years	10–77	Change in LH and FSH and T, related to age and duration of exposure. No difference in PRL.
Oldereid (1993)	41 necropsies	N/A	N/A	N/A	Measurable amounts of lead present in all reproductive organs.
Quintanilla-Vega (2000)	Human HP2 extracted from fertile donors	In vitro isolation of HP2	1 hour	N/A	Lead decreases the interaction of DNA with HP2, may alter sperm chromatin condensation, reduce fertility.

Table 6.2 (continued)

Reference[1][a]	Subjects	Route of Administration	Duration of Exposure Dosage	Blood Pb-level Measured[2][b] (µg/dl)	Results
Robins (1997)	N = 97	Lead acid battery facility	0.5–32 years	28–93	Abnormal sperm morphology, normal conc. No assoc. with fertility.
Rodamilans (1988)	Control (N = 20) Exposed (N = 23)	Worked in Pb smelting	<1 year, 1–5 yrs, >5 yrs	17–70	T decreased, LH increased in all groups, FSH did not change
Sallmen (2000)	Wives of husbands occupationally exposed (N = 515, 502)	Various occupations	5 years	10.4–39.4	Paternal exposure to lead associated with infertility, risk ratio increasingly greater with increase in exposure of lead in men.
Selevan (1984)	Wives of workers exposed (N = 376)	Through male partners	N/A	25–60	Elevated OR values for fetal loss with paternal lead exposure are observed in females who smoke during pregnancy.
Telisman (2000)	Controls (N = 51) Exposed (N = 98)	Industrial workers	2–21 years	< 10 11–70	Decrease in sperm density, counts of total motile and viable sperm, increase in abnormal sperm head morphology with increasing BLL. Serum estradiol increased. No change in T, LH, FSH, Prolactin.
Wildt (1983)	Controls (N = 31) Exposed (N = 31)	Worked in a battery factory	8 months–10 years	22–45	No diff. found in sperm morphology, count, motility, and biochemistry of all groups. Decreased sperm chromatin stability.
Xuezhi (1992)	Review of resources in China	Occupational and environmental exposure	N/A	20–60	Decrease in sperm count and motility necrospermia.

[a] First author

[b] Unless otherwise indicated

[c] Blood lead levels

In the first study to assess the physiologic site of toxicity in the reproductive axis, Braunstein et al. (1978) studied three groups of men with increasing lead levels (39μg/dl (n = 6), 29 μg/dl (n = 4), and 16 μg/dl n = 9). Decreased spermatogenesis, both in semen samples and in testicular biopsies, was noted in the lead-exposed groups. Serum testosterone levels also were suppressed in the men with the higher lead levels, and similar to the rat studies discussed, testosterone levels increased following hCG stimulation testing. However, LH response did not increase following GnRH testing, suggesting the pituitary as the site of toxicity. Others have reported a similar hormone profile consistent with a defect at the hypothalamic–pituitary level (Gustafson et al., 1989; McGregor et al., 1990). Some studies report only primary sperm toxicity or no toxicity on the HPG axis, while others report variable effects on prolactin levels (Table 6.2).

Assennato and colleagues (1987) found no differences in serum hormone levels, but a decrease in sperm concentration in a case control study. In a cross-sectional study, Alexander et al. divided men working at a lead smelter into four groups according to lead levels. Sperm concentration decreased in groups with lead levels greater than 40μg/dl, but no association was found between exposure and reproductive hormones (Alexander et al., 1996). These two studies suggest that lead is a primary germ-cell toxicant. Others reported decreased testosterone levels and increased LH levels, suggesting lead adversely impacts the testicular leydig cells as well as the germ cells (Ng et al., 1991).

Clinical studies also confirm the findings in animal studies of age-related toxicity, adaptation, and reversibility of lead's toxic effects. In a study of 122 men working in a lead battery plant, men younger than 40 years presented with different perturbations to the HPG axis than men older than 40 years (Ng et al., 1991). The older men presented with lower serum testosterone levels than younger exposed men and nonexposed controls. LH and FSH were significantly higher in exposed subjects less than 29 years old. Toxicity was related to blood lead level and duration of exposure. Subjects with at least 10 years of exposure had lower testosterone levels and tended not to respond to these lower levels with increased gonadotropins. LH and FSH levels also gradually increased with blood lead levels of 10 to 40μg/dl, and plateaued or declined at high lead levels. A few studies have focused on the impact of lead on sperm function and fertility.

In a cohort study, semen quality was investigated in 85 men employed at motorway tollgates and in 85 age-matched men living in the same area. Blood lead levels of lead were inversely correlated with sperm count and viability, but not reproductive hormone levels (De Rosa et al., 2003). As documented in animal studies, lead also disrupts the human spermatozoa's acrosome reaction. In a prospective double-blind study, sperm samples collected from the male partners of couples undergoing IVF for infertility were measured for lead content, surface receptors for mannose binding, and the ability to undergo the acrosome reaction. Lead levels in semen were negatively correlated with IVF rates, mannose receptors and mannose-induced acrosome reaction, but positively correlated with the spontaneous acrosome reaction (Benoff, 2003). Similar results were reported in *in vitro* conditions (Benoff, 2000).

Lead may also interact with calcium and/or zinc at the cellular level (Benoff et al., 2000b; Benoff et al., 2003). Lead interaction with nuclear protamines, which prevent normal sperm chromatin condensation has been proposed by Qunitanilla-Vega et al. (2000) as a mechanism of lead's toxic effect on sperm function.

Lead's impact on human fertility and fecundity is less clear cut. A cohort study conducted in a French battery factory between 1977 and 1982 divided a total of 354 battery workers into 229 lead-exposed subjects and 125 non-lead–exposed subjects. A comparison in person-year analysis did not uncover any relationship between lead exposure at any level of absorption and infertility (Coste et al., 1991). In contrast, retrospective time-to-pregnancy studies suggest that paternal lead exposure is associated with a delay in time to pregnancy (Gennert et al., 1992; Lin et al., 1996; Sallmen et al., 2000). In a case report, a firearms instructor who presented with markedly elevated levels of blood lead (880µg/dl) and oligospermia responded to chelation therapy with a dramatic increase in sperm density and a return to fertiliy with a child conceived after one year of therapy (Fisher-Fischbein et al., 1987).

Spontaneous miscarriage and congenital malformations may be related to paternal lead exposure. The odds ratio for fetal loss with paternal lead exposure may be increased (Al-Hakkak et al., 1986; Selevan et al., 1984; Lindbohm, 1991). The odds ratio for congenital malformations in the offspring of lead-exposed fathers may also be increased (Lerda, 1992; Sallmen et al., 1992).

Effects of Lead on Female Reproduction

In the late 1800s and early 1900s, small case reports described a relationship between lead exposure in women and infertility, increased rate of stillbirths, increased rate of miscarriages, and increased rates of neonatal morbidity and mortality (Putnam, 1986). There are, however, few well-controlled detailed studies evaluating the toxic effects of lead on female reproduction. In general, published studies suggest that the female is less susceptible to the disruptive effects of lead on the reproductive axis.

Animal Studies

Animal studies evaluating the toxic effects of lead on the female reproductive system are limited (Table 6.3). In a comprehensive study of the effects of *in utero* lead exposure on reproduction, Kimmel and colleagues reported a delay in vaginal opening but no effects on ability to conceive or carry normal litter to term in females chronically exposed to lead acetate throughout gestation and development (Grant et al., 1976; Kimmel et al., 1980). Others confirmed the delay in vaginal opening in the offspring of animals exposed

Table 6.3

Reference[a]	Species Strain Age at Start	Lead Compound	Route of Administration	Duration of Exposure Dosage	Blood Pb-Level at Sacrifice[b] (µg/dl)	Results
Dearth (2002)	Rat Adult Fisher 30 days prior to breeding until pups weaned at 21 days	Lead acetate	Gavage, and pups through gestation and/ or lactation	Dose: 1 ml of 12 mg/ml PbAc		Delayed timing of puberty, exhibited suppressed IGF-1, LH, and estradiol. Gestational exposure appeared to be more sensitive to lead.
Foster (1992)	Monkey Cynomolgus Newborn 300 days	Lead acetate	Gelatin capsules	300 days of life to 10 years Dose: 1500 µg/kg/day	35	Suppressed LH, FSH, E2. No overt signs of menstrual irregularity.
Hilderbrand (1973)	Rat SESCO Mature	Lead acetate	N/A	30 days Dose: 5, 100 ug	Control 14.1 Exposed 19 (for 5 µg treatment) and 30 (for 100 µg treatment)	Females treated with 5 µg showed irregularity of estrus cycle; 100 µg persistent vaginal estrus, development of ovarian follicular cysts, reduction in corpora lutea.
Kimmel (1980)	Rat CD Weanling	Lead acetate	Drinking water	6–7 weeks mated and exposed again throughout gestation and lactation Dose: 5, 25, 50, 250 ppm	Pregnancy: 0.5, 5 ppm not detectable; 25ppm = 23 µg/dl, 50 ppm= 50/35; 250 ppm= 99 µg/dl Pre-Pregnancy: 0.5, 5 ppm not detectable, 25 ppm = 19.5 µg/dl; 50 ppm = 20/23.5; 250 ppm = 31 µg/dl	Vaginal opening delayed, but did not affect ability to conceive, carry normal litter to term, or deliver the young. Percentage of malformed fetuses, resorptions, and postpartum pup deaths unaffected by Pb exposure.

Table 6.3 (continued)

Reference[a]	Species Strain Age at Start	Lead Compound	Route of Administration	Duration of Exposure Dosage	Blood Pb-Level at Sacrifice[b] (μg/dl)	Results
McGivern (1991)	Rat S-D *In utero*	Lead acetate	Drinking water	Day 14 of gestation to partuition	In weaned pups (not detectable)	Female offspring delay in vaginal opening, 50% found to exhibit prolonged / irregular periods of diestrous / absence of corpora lutea when euthenized. Pulsatile release of gonadotropins, measured in castrated animals (both sexes) revealed irregular release patterns of FSH and LH.
Piasek (1991)	Rat Albino Mature	Lead acetate	Drinking water	9 weeks (1st offspring), 20 weeks (2nd offspring) Dose: 7500 ppm	N/A	Adverse reproduction toxicity of lead is reversible if exposure is stopped
Pillai (2003)	Rat NMRI Mature	Lead acetate	Intraperitoneally	15 days	N/A	No change in LH and FSH
Pinon-Lataillade (1993)	Mouse	Lead acetate	Drinking water	Dose: 0.05 mg/kg/d 60 days	51.1–58	Exposed females mated with unexposed mice, offspring male fertility not affected, female infertility was observed: litters smaller.
Ronis (1994)	Rat S-D At birth	Lead acetate	Drinking water	85 days	N/A	Delayed female vaginal opening. Estradiol suppression to undetectable, decreases in LH.

First author[a]					BLL[b]	Effects
Ronis (1996)	Rat S-D *In utero* Prepubertally Postpubertally	Lead acetate	Drinking water	21, 35, 55, and 85 days Dose: 0, 0.6 %	0–264	Dose response delay to vaginal opening by 20 days, loss of regular estrus cycling up to 50 % of pups. Reduced circulating sex steroid levels at birth.
Ronis (1998)	Rat S-D At birth	Lead acetate	Drinking water	21, 35, 55, and 85 days Dose: 0–0.45%	0–263	Birthweight, prepubertal and pubertal growth rates significantly suppressed. Increase in pit. GH, decrease in T and estradiol observed during puberty. Postpuberty: sex steroid levels back at control levels.
Sokol (1988)	Rat Wistar 52 days	Lead acetate	Drinking water	30 days Dose: 0–0.1%	7–88	No differences in cycle length, serum LH, FSH.
Sourgens (1987)	Rat Wistar	Lead acetate	Drinking water	30 days Dose: 0–0.45%	5–60	BLL[c] 200X greater in bone, 40X in brain, and <2X ovaries, than in controls.

[a] First author
[b] Unless otherwise indicated
[c] Blood lead levels

to lead acetate *in utero* during the time of sexual differentiation, but also reported a disrupted estrous cyclicity, and absent corpora lutea (McGivern et al., 1991; Ronis et al., 1994; 1996). The latter studies reported variable effects on gonadotopins and steroid hormone levels. Female mice exposed *in utero* had reduced subsequent fertility (Pinon-Lataillade et al., 1995), but female rats did not (Coffigny et al., 1994).

Studies reporting on the toxic effects of exposure during puberty and adulthood also are inconsistent, but suggest that the animal exposed during puberty is more susceptible than the adult animal. A delay in the onset of puberty and irregular estrous cyclicity is reported in female rats treated prepubertally or during puberty (Hilderbrand, 1973; Der, 1978; Ronis et al., 1996). No change in estrous cyclicity is reported in animals treated following puberty (Sokol and Sod-Moriah, 1988; Ronis et al., 1996). Prepubertally exposed female rats have suppressed LH and estradiol levels, but adult exposed rats do not (Ronis et al., 1995; Sokol et al., 1980; Sourgens et al., 1987). Serum estradiol levels were not decreased following dosing (Sokol et al., 1988; Sourgens et al., 1987), but pituitary LH and FSH content decreased (Sourgens et al., 1987). Similar to adult dosed male rats, LH mRNA, pituitary content of LH, and GnRH mRNA increased in the adult exposed female rat (Sokol, unpublished results; Ronis et al., 1995). Also as with the male, the adverse reproductive action of lead is reversible after withdrawal of the female from exposure (Piasek et al., 1991).

Primate studies are consistent with the rodent studies. Menstrual cyclicity in monkeys was initially disrupted with lead exposure, but the animals adapted to the toxicity with normalization of cycles (Foster, 1992).

Clinical Studies

As with the animal studies, there are few studies published on the effects of lead exposure on female human reproduction (Table 6.4). Two recent epidemiologic studies suggest that lead delays puberty in girls in a fashion similar to that reported in the animal studies. Both studies analyzed data from the Third National Health and Nutrition Examination Survey (NHANES III, 1988–94). The NHANES III was a cross-sectional survey that used a stratified multistage probability sampling design to obtain nationally representative information on the health and nutrition of the U.S. population (Pirkle et al., 1998; Wu et al., 2003). Within the total sample survey of 39,695 individuals, 7050 were girls between 1 and 16 years of age. These studies analyzed data regarding menstrual history and Tanner staging of pubic hair and breast development collected on 1706 girls 8 to 16 years old who had documented lead levels. SES and race were self-reported. Both studies reported that higher blood lead levels were significantly associated with delayed attainment of menarche and pubic hair, regardless of SES or ethnicity (Wu et al., 2003; Selevan et al., 2003). Delay in breast development was reported to occur more commonly in African-American girls (Selevan et al., 2003).

Table 6.4

Reference[a]	Subjects	Route of Administration	Duration of Exposure Dosage	Blood Pb-level Measured[b] (μg/dl)	Results
ATDSR (1997)	Women exposed at a mining and smelter complex	Environmental and occupational	At least 30 days	0–100 μg/g bone mineral	Early menopause, increased incidence of osteoporosis.
Gulson (1999)	Pregnant women (N = 28) 23 migrant women to Australia and 5 Australian born	Food	6-day study Dosage: 5.8 ± 3 Range was 2–39 μg/kg/day	Mean BLL[c] = 3	Dietary lead intake low: skeletal contribution is the dominant contributor to blood lead, esp. during pregnancy and postpartum.
Sallmen (1995)	(N = 121)	Occupationally exposed	N/A	10–50	No diff. In distribution of time to pregnancy between exposed and nonexposed
Selevan (2003)	Girls 8–18 years of age (N = 2186)	Environmental lead	N/A	1–10	Decreased height; significant delays in breast and pubic-hair development in African-American and Mexican-American girls; associated delay in age at menarche was 3.6 months
Wu (2003)	U.S. girls between 8–16 years old (N = 1706)	Environmental or community exposure	N/A	0.7–21.7	Higher blood lead levels were significantly associated with delayed achievement of menarche and pubic hair among U.S. girls, but not with breast development
Xuezhi (1992)	Review of research in China 1978–1991	Occupational and environmental	N/A	N/A	Irregular menstrual cycle, increase of spontaneous abortions

a First author
b Unless otherwise indicated
c Blood lead levels

Early studies suggested that lead exposure was associated with decreased fertility (Rom, 1976). A small retrospective time-to-pregnancy study of lead-exposed women did not support the hypothesis that maternal exposure to inorganic lead at low exposure levels is associated with reduced fertility (Sallmen et al., 1995). However, the participation rate was very small and the results must be considered as preliminary.

Early menopause may be associated with lead exposure. The Agency for Toxic Substances and Disease Registry Study (ATDSR, 1997) reported findings from a self-reported questionnaire regarding reproductive outcomes in women exposed occupationally to lead. Exposed women were more likely to have stopped having periods at a statistically younger age, and were more likely to be diagnosed with osteoporosis than did nonexposed women.

The issue of bone stores of lead is particularly important when discussing the toxic effects of lead on the female. Pregnancy, lactation, and menopause are times of high bone turnover, and accumulated bone lead stores may constitute an internal source of exposure (Hernandez-Avila, 1998). Accumulated bone lead constitutes a moderate source of circulating lead during adolescence (Farias et al., 1998). Changes in blood lead levels and isotopic composition could not be explained by dietary lead in studies evaluating the impact of diet on lead in blood and urine of pregnant or lactating women (Gulson et al., 1998), suggesting bone stores as the source of the lead.

Summary

Animal and clinical studies support the hypothesis that lead exposure disrupts male reproduction. Toxicity is age and dose related, adaptable, and reversible. Toxicity appears to involve multiple sites within the hypothalamic-pituitary-testicular axis. The exact mechanisms of this toxicity are currently being defined.

Data generated in females are less consistent, although delays in pubertal milestones have been documented in animals and humans. Questions remain as to the impact of lead exposure on fertility and menopause.

References

Agency for Toxic Substances and Disease Registry (1996). *Report of Study of female former workers at a lead smelter: Effects of lead exposure on bone density and other health outcomes.* Atlanta, Georgia, U.S. Dept. of Health and Human Services, Public Health Service.

Alexander, B., Checkoway, H., et al. (1996). "Semen quality of men employed at a lead smelter." *Occup Environ Med* **53**: 411–416.

Al-Hakkak, Z. (1988). "Effects of Ingestion of lead monoxide alloy on male mouse." *Arch Toxicol* **62**: 97–100.

Al-Hakkak, Z., Hamamy, H. et al. (1986). "Chromosome aberrations in workers at a storage battery plant in Iraq." *Mutat. Res* **171**: 53–60.

Apostoli, P., et al (1998). "Male Reproductive Toxicity of Lead in Animals and Humans." *Occup Environ Med* **55**: 364–374.

Apostoli, P., Porru, S., et al. (1999). "Critical aspects of male fertility in the assessment of exposure to lead." *Scand. J. Work Environ. Health* **25**(Suppl. 1): 40–43.

Assennato, G., et al (1987). "Sperm Count Suppression without Endocrine Dysfunction in Lead-Exposed Men." *Arch Environ Health* **42**(2): 124–127.

Assennato, G., Paci, A., et al. (1986). "Sperm count suppression without endocrine dysfunction in lead-exposed men." *Arch. Environ. Health* **41**: 387–390.

Barraclough, C. (1982). "The role of catecholamines in the regulation of pituitary LH and FSH secretion." *Endocr Rev* **3**: 91.

Barratt, C., Davies, A., Bansal, M., Williams, M. (1989). "The effect of lead on the male reproductive system." *Andrologia* **21**: 161–166.

Batra, N., Nehru, B., et al. (1998). "The effect of zinc supplementation on the effects of lead on the rat testis." *Reproductive Toxicology* **12**(5): 535–540.

Bechter, R., Ettlin, R., Dixon, R. (1982). "Assessment of the testicular toxicity associated with anti-cancer agents.II Sperm counts and serial mating." *Proc West Pharmacol Soc* **25**: 385–387.

Bell, J., and Thomas, J. (1980). Effects of lead on Mammalian Reproduction. *Lead Toxicity.* R. Singhal, Thomas, JA., eds., Urban & Schwarzenberg, Baltimore-Munich: 169–183.

Benoff, S. (2000). "Male infertility and environmental exposure to lead and cadmium." *Human Reproduction Update* **6**(2): 107–121.

Benoff, S., Cooper, G., Jacob, A., Hershlag, A., Hurley, I. (2000). "Metal ions and human sperm mannose receptors." *Andrologia* **32**(4–5): 317–29.

Benoff, S., Jacob, A., Hurley, I., et al (2000). *Human Reprod Update* **6**(2): 107–121.

Benoff, S., Millan, C., et al. (2003). "The Role of Zinc (Zn) in Lead (Pb)-Induced Reproductive Toxicity." *In press.*

Boscolo, P. (1988). "Ultrastructure of the testis in rats with blood hypertension induced by long term lead exposure." *Toxicology* **41**: 129–137.

Braunstein, G., Dahlgren, G., et al. (1978). "Hypogonadism in Chronically Lead - Poisoned Men." *Infertility* **1**: 33–51.

Chapin, R., Filler, R. et al. (1992). "Methods for Assessing Rat Sperm Motility." *Reproductive Toxicology* **6**(3): 267–273, 34 references.

Chowdhuri, D., Narayan, R., et al. (2001). "Effect of lead and chromium on nucleic acid and protein synthesis during sperm-zona binding in mice." *Toxicology in Vitro* **15**(6): 605–613.

Chowdhury, A. (1986). "Effect of lead in human semen." *Adv. Contracept Delivery Syst* **2**: 208–211.

Chowdhury, A., et al (1984). "Toxic effect of lead on the testes of rat." *Biomed Biochem Acta* **43**(1): 95–100.

Chowdhury, A., et al (1986). "Histochemical changes in the testes of lead induced experimental rats." *Folia Histochem et Cytobiol* **24**: 233–238.

Clermont, Y. (1986). "The cycle of seminiferous epithelium in man." *Am J Anat* **112**: 35.

Coffigny, H., et al. (1994). "Effects of lead poisoning of rats during pregnancy on the reproductive system and fertility of their offspring." *Human & Experimental Toxicology* **13**(4): 241–246.

Consumer Lab (2003). Product Review: Calcium, Consumer Lab. **2003**.

Corpas, I., Antonio, M. (1998). "Study of alterations produced by cadmium and cadmium/lead administration during gestational and early lactation periods in the reproductive organs of the rat." *Ecotoxicol Environ Saf* **41**(2): 180–188.

Coste, J., Mandereau, P., Bregu, M., Faye, C., Hemon, D., Spira, A. (1991). "Lead-exposed workmen and fertility: a cohort study on 354 subjects." *Eur J Epidemiol* **7**(2): 154–158.

Cullen M., Robins, J., et al (1984). "Endocrine and Reproductive Dysfunction in Men Associated with Occupational Inorganic Lead Intoxication." *Arch Environ Health* **39**: 421–440.

Cunningham, M. (1986). "Chronic Occupational Lead Exposure: the Potential Effect on Sexual Function and Reproductive Ability in Male Workers." *Occupational Health Nursing* **34**(6): 277–279.

De Rosa, M., Zarrilli, S., et al. (2003). "Traffic pollutants affect fertility in men." *Human Reproduction* **18**(5): 1055–1061.

Dearth, R., Hiney, J., Srivastava, V., Burdick, S., Bratton, G., Dees, W. (2002). "Effects of lead (Pb) exposure during gestation and lactation on female pubertal develoment in the rat." *Reprod Toxicol* **16**(4): 343–352.

Der, R., Fahim, Z., Yousef, M., Fahim, M. (1976). "Environmental interaction of lead and cadmium on reproduction and metabolism of male rats." *Res Commun Chem Pathol Pharmacol* **14**: 689–713.

Eyden, B., Maisin, J., et al. (1978). "Long-Term Effects Of Dietary Lead Acetate On Survival, Body Weight And Seminal Cytology In Mice." *Bull Environ Contam Toxicol* **19**: 266–272.

Factor-Litvak, P., Graziano, J., et al. (1991). "A Prospective Study of Birthweight and Length of Gestation in a Population Surrounding a Lead Smelter in Kosovo, Yugoslavia." *International Journal of Epidemiology* **20**(3): 722–728.

Farias, P. (1998). "Determinants of Bone and Blood Lead Levels among Teenagers Living in Urban Areas with High Lead Exposure." *Environ Health Perspect* **106**: 733–737.

Fisch, H. (1996). "Semen analyses in 1283 men from the United States over a 25-year period: no decline in quality." *Fertility and Sterility* **65**(5): 1009–1014.

Fisher-Fischbein, J., Fischbein, H., Meinick, C., Bardin, W. (1987). "Correlation between biochemical indicators of lead exposure and semen quality in a lead-poisoned firearms instructor." *JAMA* **257**: 803–805.

Foote, R. (1999). "Fertility of rabbit sperm exposed in vitro to cadmium and lead." *Reproductive Toxicology* **13**(Issue 6): 443–449.

Foster, W. (1992). "Reproductive toxicity of chronic lead exposure in the female Cynomolgus monkey." *Reprod Toxicol* **6**: 123–131.

Foster, W., McMahon, A., et al. (1993). "Reproductive Endocrine Effects of Chronic Lead Exposure in the Male Cynomolgus Monkey." *Reproductive Toxicology* **7**(3): 203–209, 31 references.

Fowler, B., Kimmel, C., et al. (1980). "Chronic low-level lead toxicity in the rat: 3. An integrated assessment of long-term toxicity with special reference to the kidney." *Toxicol Appl Pharmacol.* **56**(1): 59–77.

Gennart, J., Bernard, A., et al. (1992). "Assessment of Thyroid, Testes, Kidney and Autonomic Nervous System Function in Lead-Exposed Workers." *International Archives of Occupational and Environmental Health* **64**(1): 49–57.

Gennart, J., Buchet, J., Roels, H., Ghyselen, P., Ceulemans, E., Lauwerys, R. (1992). "Fertility of male workers exposed to cadmium, lead, or manganese." *Am. J. Epidemiol* **135**: 1208–1219.

Gerhard, I. (1993). "Reproductive Risks of Heavy Metals and Pesticides in Women." *Reproductive Toxicology*: 167–183.

Gilifillan, S., et al (1965). "Lead poisoning and the fall of Rome." *J Occup Med* **7**: 53–60.

Govoni, S., Battaini, F., Fernicola, C., et al (1987). "Plasma prolactin concentrations in lead exposed workers." *J Environ Path Toxicol Oncol* **7**: 13–15.

Grant, L., et al (1976). "Abstr. 1640." *Fedn Proc* **35**: 503.

Grant, L., et al (1980). "Chronic low level lead toxicity in the rat." *Toxicol Appl Pharmacol* **56**: 42–58.

Gulson, B., et al (1998). "Relationships of lead in breast milk to lead in blood, urine, and diet of the infant and mother." *Environ Health Perspect* **106**(10): 667–674.

Gulson, B., et al (1999). "Impact of diet on lead in blood and urine in female adults and relevance to mobilization of lead from bone stores." *Environ Health Perspect* **107**(4): 257–263.

Gustafson, A., Hedner, P., et al. (1989). "Occupational Lead Exposure and Pituitary Function." *Int Arch Occup Environ Health* **61**(4): 277–282.

Health Council of the Netherlands: Committee for Compounds Toxic to Reproduction (2003). Metallic Lead; Evaluation of the Effects of Reproduction, Recommendation for Classification., The Hague: Health Council of the Netherlands. **2003**.

Hernández-Avila, M. (1998). "The Influence of Bone and Blood Lead on Plasma Lead Levels in Environmentally Exposed Adults." *Environmental Health Perspectives* **106**(8): 473–477.

Hernández-Serrato, M., Mendoza-Alvarado, L., Rojas-Martinez, R., Gonzalez-Garza, C., Hulme, J., Olaiz-Fernandez, G. (2003). "Factors associated with lead exposure in Oaxaca, Mexico. " *Journal of Exposre Analysis and Environmental Epidemiology* **13**(5): 341–347.

Hibbert, R., Navia, B., Kammen, D., Zhang, J. (1999). "High lead exposures resulting from pottery production in a village in Michoacan State, Mexico." *Expo Anal Environ Epidemiol.* **9**(4): 343–351.

Hilderbrand, D., Der, R., et al. (1973). "Effect of lead acetate on reproduction." *Am J Obstet Gynecol* **115**(8): 1058–1065.

Hoff, J., Quigley, M., Yen S. (1983). "Hormonal dynamics at midcycle: a reevaluation." *J Clin Endocrinol Metab* **57**:792.

Hopkins, R., Minnich, S., et al. (1991). "Maternal occupational exposures during pregnancy and risk of preterm delivery." *Am J Epidemiol* **134**(7): 723.

Hsu, P., Hsu, C., et al. (1998). "Lead-induced changes in spermatozoa function and metabolism." *J Toxicol Environ Health A.* **55**(1): 45–64.

Ivanova-Chemishanska, L., et al. (1980). "Multigeneration studies of white rats obtained from male parent generation, exposed to lead acetate." *Toxicol Lett* **1000**: 111.

Johansson, L. (1989). "Premature acrosome reaction in spermatazoa from lead-exposed mice." *Toxicology* **54**: 151–162.

Johansson, L., et al. (1987). "Effects of Lead on the Male Mouse as Investigated by In Vitro Fertilization and Blastocyst Culture." *Environmental Research* **42**(1): 140–148, 24 reference.

Johansson, L., Wade, M. (1986). "Long term exposure of the male mouse to lead: Effects on fertility." *Environ Res* **41**: 481–487.

Kempinas, W., et al (1994). "Time-dependant effects of lead on rat reproductive functions." *Journal of Applied Toxicol* **14**(6): 427–433.

Kimmel, C., Grant, L., et al (1980). "Chronic Low level dose toxicity in the rat I. Maternal Toxicity and Perinatal effects." *Toxicol Appl Pharmacol.* **56**: 28–41.

Klages (1987). "Gonadal and thyroid function after experimental lead exposure." *Trace Elements in Medicine* **4**(1): 8–12.

Klein, D., Wan, Y., et al. (1994). "Effects of toxic levels of lead on gene regulation in the male axis: increase in messenger ribonucleic acids and intracellular stores of gonadotrophs within the central nervous system." *Biol Reprod* **50**(5): 802–811.

Krasovskii, G., Vasukovich, L., et al (1979). "Experimental study of biological effects of lead and aluminum following oral administration." *Environ Health Perspect* **30**: 47–51.

Lancranjan I and P. H. e. al (1975). "Reproductive ability of workmen occupationally exposed to lead." *Arch Environ Health* **30**: 396–401.

Lanphear, B., Canfield, R., et al. (2001). "Environmental exposure to lead and children's intelligence at blood lead concentrations below 10 micrograms per deciliter." *Pediatr Res* **49**: 16A.

Leggette, R. (1993). "An age-specific kinetic model of lead metabolism in humans." *Environ Health Perspect* **101**: 598–616.

Lerda, D. (1992). "Study of sperm characteristics in persons occupationally exposed to lead." *Am J Ind Med* **22**(4): 567–571.

Li, S. (1993). "Reproductive toxicology - China." *Reproductive Toxicology*: 63–71.

Lin, S., Hwang, S., et al (1996). "Fertility rates among lead workers and professional bus drivers: a comparative study." *Ann Epidemiol* **6**: 201–208.

Lindbohm, M. (1991). "Paternal occupational lead exposure and spontaneous abortion." *Scand. J. Work Environ. Health* **17**: 95–103.

Lobo, R. (1997). "The menstrual cycle." in *Infertility, Contraception, and Reproductive Endocrinology*, R. Lobo, D. Mishell, Jr., R. Paulson and D. Shoupe, eds., Blackwell Science, Inc., Malden, Massachusetts, 124–140.

Mann, T. (1981). "Male Infertility." *Metheum and Co., Ltd, London.*

Mann, T., Lutwak-Mann, C. (1981). *Male Reproductive Function and Semen*. Berlin, Heidelberg, New York, Springer-Verlag.

Martäinez, S., et al. (1994). "Effects of prenatal administration of low-concentrations of lead on newborn rats." *Revista de Toxicologia* **11**(1): 16–19.

Matsumoto, A., Bremner, W. (1987). "Endocrinology of the HPT axis with the particular reference to the hormonal control of spermatogenesis." *J Clin Endocrinol Metab* **2**: 103.

McGivern, R., et al (1991). "Prenatal Lead Exposure in the Rat during the Third Week of Gestation: Long-Term Behavioral, Physical, and Anatomical Effects Associated with Reproduction." *Toxicol Appl Pharmacol.* **110**: 206–215.

McGregor, A., Mason, H. (1990). "Chronic occupational lead exposure and testicular endocrine function." *Hum Exp Toxicol.* **9**: 371–376.

Millan, C., Sokol, R., et al. (2003). "Lead Induces Epigenetic Modification of Rat Testicular Gene Expression: A DNA Microarray Study." *in press.*

Mishell, D., Jr., Nakamura, R., et al (1971). "Serum gonadotropin and steroid patterns during the normal menstrual cycle." *Am J Obstet Gynecol* **111**: 60.

Moorman, W., Clark, J., et al (1995). "Validation of the rabbit model for assessing reproductive toxicants. High dose phase report." *NIOSH report.*

Nathan, E., et al (1992). "Lead Acetate does not Impair Secretion of Sertoli Cell Function Marker Proteins in the Adult Sprague Dawley Rat." *Arch Environ Health* **47**(5): 370–375.

National Health and Nutrition Examination Survey III, e. a. (1988–94). "NHANES III series II." *National Center for Health Statistics.*

Needleman, H., et al. (1990). "The long-term effects of exposure to low doses of lead in childhood: An 11-year follow-up report." *N Engl J Med* **322**(2): 83–88.

Ng, T., Goh, H., et al. (1991). "Male Endocrine Functions in Workers with Moderate Exposure to Lead." *British Journal of Industrial Medicine* **48**(7): 485–491.

Oldereid, N., Thomassen, Y., et al. (1993). "Concentrations of lead, cadmium and zinc in the tissues of reproductive organs of men." *J Reprod Fertil* **99**(2): 421–425.

Petrusz, P., Weaver, G., Grant, L., et al (1978). "Lead poisoning and reproduction: effects on pituitary and serum gonadotropins in neonatal rats." *Environ Res* **19**: 383–391.

Piasek, M., et al. (1991). "Reversibility of the effects of lead on the reproductive performance of female rats." *Reprod Toxicol* **5**(1): 45–51.

Piasek, M., et al. (1987). "Effect of exposure to lead on reproduction in male rats." *Bull Environ Contam Toxicol* **39**: 448–452.

Pillai, A., Priya, L., Gupta, S. (2003). "Effects of combined exposure to lead and cadmium on the hypothalamic-pituitary axis function in proestrous rats." *Food and Chemical Toxicology* **41**(3): 379–84.

Pinon-Lataillade, G., et al (1993). "Effect of ingestion and inhalation of lead on the reproductive system and fertility of adult male rats and their progeny." *Hum Exp Toxicol.* **12**: 165–172.

Pinon-Lataillade, G., Thoreux-Manlay, A., et al. (1995). "Reproductive toxicity of chronic lead exposure in male and female mice." *Hum Exp Toxicol.* **14**(11): 872–878.

Pirkle, J., et al. (1998). "Exposure of the U.S. population to lead, 1991–1994." *Environ Health Perspect* **106**(11): 745–750.

Plant, T. (1986). "Gonadal regulation of hypothalamic gonadotropin releasing hormone in primates." *Endocr Rev* **7**: 75.

Polato, R., Morossi, G., Furlan, I., Moro, G. (1989). "Risk of abnormal lead absorption in glass decoration workers." *Med Lav* **80**(2): 136–139.

Prevention, C. f. D. C. a. (2003). "Second National Report on Human Exposure to Environmental Chemicals."

Puhac, I., et al (1963). "Labratory investigation on the possibility of employing lead compounds as raticides by decreasing the reproductive capacity of rats." *Acta Vet* **13**: 3–9.

Putnam, R. (1986). "Review of Toxicology of Inorganic Lead." *Am Ind Hyg Assoc* **47**(11): 700–703.

Quintanilla-Vega B., Bal W., Silbergeld, E., Waalkes, M., Anderson, L. (2000). "Lead interaction with human protamine (HP2) as a mechanism of male reproductive toxicity." *Chem Res Toxicol* **13**(7): 594–600.

Rabinowitz, M., Wetherill, G., Kopple, J. (1973). "Assessing Human Lead Metabolism by Satble Isotope Techniques." *Fedn Proc* **32**(3 part 1): 930.

Robins, T., Bornman, M., Ehrlich, R., Cantrell, A., Pienaar, E., Vallabjee, J., Miller, S. (1997). "Semen quality and fertility of men employed in a South African lead acid battery plant." *American Journal of Industrial Medicine* **32**: 369–376.

Rodamilans, M., Osaba, M., To-Figueras, J., Fillat, F., Corbella, J. (1988). "Inhibition of intratesticular testosterone synthesis by organic lead." *Toxicology Letters* **42**: 285–290.

Rodamilans, M., Osaba, M., To-Figueras, J., Fillat, F., Marques, J., Corbella, J. (1988). "Lead toxicity on endocrine testicular function in a occupationally exposed population." *Hum Toxicol* **7**: 125–128.

Rom, W. (1976). "Effects of lead on the female and reproduction." *Mt Sinai J Med* **43**: 542–546.

Ronis, M., Bell, L., et al. (1996). "Neuroendocrine mechanisms underlying the reproductive toxicity of lead in a rat lifetime exposure model." *Metal Ions in Biology and Medicine* **4**: 420–422.

Ronis, M., et al (1996). "Reproductive Toxicity and growth effects in rats exposed to lead at different periods during development." *Toxicol Appl Pharmacol.* **136**: 361–371.

Ronis, M., Gandy, J., et al. (1998). "Endocrine mechanisms underlying reproductive toxicity in the developing rat chronically exposed to dietary lead." *J Toxicol Environ Health A* **54**(2): 77–99.

Ronis, M., Shahare, M., et al. (1994). "Reproductive toxicity and disrupted pubertal growth in rats exposed to lead during different developmental periods." *Toxicologist* **14**(1): 84.

Ronis, M., et al (1997). "Effects on pubertal growth and reproduction in rats exposed to lead perinatally or continuously throughout development." *J Toxicol Environ Health* **53**: 327–341.

Ross, E., Szabo, N., Tebbett, I (2000). "Lead content of calcium supplements. " *JAMA* **284**:1425–1429.

Sallmén, M., et al. (1995). "Time to Pregnancy among Women Occupationally Exposed to Lead." *Journal of Occupational and Environmental Medicine* **37**(8): 931–934.

Sallmén, M., et al. (1992). "Paternal Occupational Lead Exposure and Congenital Malformations." *Journal of Epidemiology and Community Health* **46**(5): 519–522.

Sallmén, M., Lindbohm, M., Nurminen, M. (2000.). " Paternal exposure to lead in infertility." *Epidemiology* **11**: 148–152.

Saxena, D. (1984). "Lead Induced Histochemical Changes in the Testes of Rats." *Industrial Health* **22**: 255–260.

Saxena, D. (1986). "Lead Induced Testicular Dysfunction in Weaned Rats." *Industrial Health* **24**: 105–109.

Saxena, D., et al (1987). "The effect of lead exposure on the testis of growing rats." *Exp Pathol* **31**: 249–252.

Schrader, S., Kanitz, H. (1994). "Occupational Hazards to Male Reproduction." *Occupational Medicine: State of the Art reviews*: 405–414.

Schroeder, H., Mitchner, M. (1971). "Toxic effects of trace elements on the reproduction of mice and rats." *Arch Environ Health* **23**: 102–106.

Selevan, S., et al. (1984). "Reproductive Outcomes In Wives Of Lead Exposed Workers." *NIOSH, U.S. Department of Health and Human Services, Cincinnati, Ohio*: 1–44.

Selevan, S., Rice, D., Hogan, K., Euling, S., Pfahles-Hutchens, M., Bethel, J. (2003). "Blood Lead Concentration and Delayed Puberty in Girls." *N Engl J Med* **348**(1): 527–1536.

Sheiner, E., Sheiner, E. (2003). "Effect of Occupational Exposures on Male Fertility: Literature Review." *Industrial Health* **41**: 55–62.

Shoupe, D., Lobo, R. (1997) "Reproductive Neuroendocrinology," in *Infertility, Contraception, and Reproductive Endocrinology*, R. Lobo, D. Mishell, Jr., R. Paulson and D. Shoupe, eds., Blackwell Science, Inc., Malden, Massachusetts, 3–25.

Silbergeld, E. (1991). "Lead in bone: implications for toxicology during pregnancy and lactation." *Environ Health Perspect* **91**: 63.

Silbergeld, E., Schwartz, J., Mahaffey, K. (1988). "Lead and osteoporosis: mobilization of lead from bone in postmenopausal women." *Environ Res* **47**: 79–94.

Sokol, R. (1987). "Hormonal effects of lead acetate in the male rat: mechanism of action." *Biol Reprod* **37**(5): 1135–1138.

Sokol, R. (1988). "Endocrine Evaluation in the Assessment of Male Reproductive Hazards." *Reproductive Toxicology* **2**: 217–222.

Sokol, R., Sod-Moriah, U. (1988). "Lead exposure and the hypothalamic-pituitary-ovarian axis in pubertal rats. " *Infertility* **11**(4): 265–271.

Sokol, R. (1989). "Reversibility of the Toxic Effect of Lead on the Male Reproductive Axis." *Reproductive Toxicology* 3(3): 175–180, 39 references.

Sokol, R., et al. (1994). "Lead exposure in-vivo alters the fertility potential of sperm in-vitro." *Toxicol Appl Pharmacol.* **124**(2): 310–316.

Sokol, R. (1997). "Male factor in infertility." in *Infertility, Contraception, and Reproductive Endocrinology,* R. Lobo, D. Mishell, Jr., R. Paulson and D. Shoupe, eds., Blackwell Science, Inc., Malden, Massachusetts, 547566.

Sokol, R., Berman, N. (1991). "The Effect of Age of Exposure on Lead-Induced Testicular Toxicity." *Toxicology* **69**(3): 269–278, 24 references.

Sokol, R., Berman, N., et al. (1998). "Effects of Lead Exposure on GnRH and LH Secretion In Male Rats: Response to Castration and -methyl-p-tyrosine (AMPT) Challenge." *Reproductive Toxicology* **12**(3): 347–355.

Sokol, R., Madding, C., et al. (1985). "Lead toxicity and the hypothalamic-pituitary-testicular axis." *Biology of Reproduction* **33**: 722–728.

Sokol, R., Sod-Moriah, U. (1988). "Lead exposure and hypothalamic-pituitary-ovarian axis in pubertal rats." *Infertility* 11(4): 265–271.

Sokol, R. (1998). Lead Neuroendocrine Toxicity. *Reproductive and Developmental Toxicology.* K. Korach. Research Triangle Park, North Carolina, Marcel Dekker INC: 249–257.

Sokol, R., et al (1990). "The Effect of Duration of Exposure on the Expression of Lead Toxicity on the Male Reproductive Axis." *Journal of Andrology* **11**(6): 521–526.

Sokol, R., Wang, S., et al. (2002). "Long-Term, Low-Dose Lead Exposure Alters the Gonadotropin-Releasing Hormone System in the Male Rat." *Environmental Health Perspectives* **110**(9): 871–874.

Sourgens, H., et al (1987). "Gonadal and Thyroid Function after Experimental Lead Exposure." *Trace Elements in Medicine* **4**(1): 8–12.

Symanski, E., Hertz-Picciotto, I. (1995). "Blood lead levels in relation to menopause, smoking, and pregnancy history." *Am J Epidemiol* **141**(11): 1047–1058.

Telisman, S., Cvitkovic, P., et al. (2000). "Semen Quality and Reproductive Endocrine Function in Relation to Biomarkers of Lead, Cadmium, Zinc, and Copper in Men." *Environmental Health Perspectives ,* **108**(1): 45–53.

Thoreux-Manlay, A., et al. (1995). "Impairment of Testicular Endocrine Function after Lead Intoxication in the Adult Rat." *Toxicology* **100**(1–3): 101–109, 43 references.

Todd, A., Wetmur, J., et al. (1996). "Unraveling the Chronic Toxicity of Lead: An Essential Priority for Environmental Health." *Environmental Health Perspectives* **104**(Supplement 1): 141–146.

Vahter, M., Berglund, M., Akesson, A. (2004). "Toxic metals and the menopause. " *J Br Menopause Soc* **10**(2): 60–64.

Varma, M., et al. (1974). "Mutagenicity and Infertility Following Administration of Lead Sub-Acetate To Swiss Male Mice." *Experientia* **30**: 486–487.

Veeramachaneni, D., Palmer, J., et al. (2001). "Long-term effects on male reproduction of early exposure to common chemical contaminants in drinking water." *Human Reprod* **16**: 979–987.

Vermande-Van Ecke, G., et al (1960). "Changes in the ovary of the rhesus monkey after chronic lead intoxication." *Fertility and Sterility* **11**(2): 223–234.

Wadi, S., Ahmad, G. (1999). "Effects of lead on the male reproductive system in mice." *J Toxicol Environ Health A.* **56**(7): 513–521.

Watson, L., et al. (1993). "Lead: interactions of pregnancy and lactation with lead exposure in rats." *Toxicologist* **13**(1): 349.

Wiebe, J., et al (1982). "Lead administration during pregnancy and lactation affects steroidogenesis and hormone receptors in testes of offspring." *J Toxicol Environ Health* **10**: 653–666.

Wiebe, J., et al (1983). "On the mechanism of action of lead in the testes: in vitro suppression of FSH receptors, cyclic AMP and steroidogenesis." *Life Sciences* **32**: 1997–2005.

Wildt, K., Eliasson, R., Berlin, M. (1983). Effects of occupational to lead on sperm and semen. *Reproductive and developmental toxicity of metals.* T. Clarkson, Nordberg, GF, Sager, P. New York, Plenum Press: 279–300.

Willems (1982). "Absence of an effect of lead acetate on sperm morphology, sister chromatid exchanges or on micronuclei formation in rabbits." *Arch Toxicol* **50**: 149–157.

Winder, C. (1989). "Reproductive and Chromosomal Effects of Occupational Exposure to Lead in the Male." *Reproductive Toxicology* **3**(4): 221–233.

Winder, C. (1993.). "Lead, reproduction and development." *Neurotoxicology* **14**(2–3): 303–317.

Wu, T., Buck, G. (2003). "Blood Lead Levels and Sexual Maturation in U.S. Girls: The Third National Health and Nutrition Examination Survey, 1988–1994." *Environmental Health Perspectives* **11**(5).

Wyrobeck, A., Bruce, W. (1978). "The introduction of sperm abnormalities in mice and humans." *Chem Mutagens* **2**: 257–258.

Xuezhi, J. (1992). "Studies of lead exposure on reproductive system: a review of work in China." *Biomed Environ Sci* **5**: 266–275.

Zirkin, B., et al (1985). "Effects of Lead Acetate on Male Rat Reproduction." *Concepts in Toxicology* **3**: 138–145.

chapter 7

Impact of Metals on Ovarian Function

Patricia B. Hoyer
Department of Physiology, University of Arizona, Tucson

Contents

Introduction

With the trend toward women starting families later in life, the effects of long-term environmental exposures on fertility must be considered. The world experienced rapid technological and industrial advancement during the last part of the 20th century, making it increasingly important to design and improve approaches for identifying environmental factors that are potentially detrimental to reproductive function in women. One class of environmental contaminants that has not been extensively studied to date is the trace metals. It is generally assumed that trace metals in high concentrations in the environment have the potential to adversely impact health in humans as well as wildlife populations. There are examples of health risks in humans known to be associated with trace metal exposures; however, the impact of these agents on reproductive health in women has not been studied in depth at a comprehensive and mechanistic level. Based on findings in animal studies, as well as epidemiological information on women, there is potential for disruption of ovarian function by the trace metals cadmium (Cd), lead (Pb), mercury (Hg), chromium (Cr), and arsenic (As). To date, no comprehensive review of the effects of these metals on ovarian function has been available. This chapter attempts to provide a compilation of the literature with respect to how these metals might affect the ovary. Therefore, the remainder of this chapter will focus on the potential for exposure to these metals to impact reproduction in females by impairing ovarian function. In order for the reader to fully appreciate the possible consequences of trace-element exposures, the chapter begins with an overview of ovarian function and follicular development (summarized from Hoyer and Devine, 2001; Hirshfield, 1991; Richards et al., 1987).

Ovarian Physiology

Endocrine Regulation

The major functions of the ovary are production of the female germ cell, the oocyte, and production of female sex steroid hormones. The ovary of the mature mammalian female contains a heterogeneous mixture of structures that undergo dynamic changes during the estrous/menstrual cycle. Specifically, the ovary contains two endocrine glands, the follicle and the corpus luteum. The follicle is responsible for gametogenesis (oogenesis) and production of the hormone 17β-estradiol (steroidogenesis). Synthesis of 17β-estradiol is regulated by the gonadotropins follicle stimulating hormone (FSH) and luteinizing hormone (LH) secreted by the anterior pituitary. The other endocrine system, the corpus luteum, is derived from follicular tissue following ovulation and is present on the ovary of the nonpregnant female only during the second half of the estrous/menstrual cycle, but is maintained if pregnancy occurs. The corpus luteum produces the steroid hormone progesterone, which is under regulation by LH.

17β-Estradiol and progesterone comprise the major ovarian steroids. 17β-Estradiol is responsible for follicular maturation, hyperplasia of the endometrium, and uterine vasculature during the follicular phase. Additionally, 17β-estradiol provides both negative and positive feedback on the hypothalamus and pituitary for regulation of secretion of the gonadotropins, LH and FSH. Progesterone facilitates implantation by preparing the uterus to accommodate a blastocyst, and provides maintenance of pregnancy by inhibiting uterine contractions and endometrial sloughing. Additionally, progesterone provides negative feedback on the hypothalamus and pituitary for inhibition of release of the gonadotropins.

Tonic secretion of LH and FSH from the anterior pituitary can be affected by the inhibitory actions of 17β-estradiol and progesterone in a classical long loop of negative feedback. Because both steroids participate in the inhibition of gonadotropin release, the absence of 17β-estradiol, progesterone, or both causes basal LH and FSH levels to increase. Additionally, inhibin produced by granulosa cells in developing follicles contributes to the selective inhibition of FSH release. In contrast to tonic LH secretion, which is controlled by a negative feedback mechanism, an LH surge that triggers ovulation is produced by a neuroendocrine reflex arc of positive feedback, which is stimulated by increasing, marked elevations in circulating 17β-estradiol levels that are produced by the largest developing preovulatory (Graafian) follicle. As a result, when levels become sufficiently elevated, 17β-estradiol shifts from its inhibitory role in regulating LH and FSH release, and initiates the LH surge (which triggers ovulation). In contrast to 17β-estradiol, progesterone inhibits the LH surge. This ensures that estrogen output in early pregnancy does not stimulate ovulation.

Follicular Development

At birth, the ovary contains its lifetime complement of germ cells because oogonia are only formed during fetal development. Oogonia become oocytes when they enter the prophase of the first meiotic division and are incorporated into primordial follicles. A primordial follicle consists of an oocyte surrounded by a complete layer of granulosa-like cells. The oocyte, which is arrested in an early stage of meiosis, remains in this state of suspended animation until the follicle receives a signal for activation of development. In the mature ovary, once a follicle begins to develop, it is committed to one of two fates. If the hormonal milieu is optimal, it will continue to develop and ovulate. Alternatively, the follicle may degenerate by a process of physiological cell death (apoptosis) called atresia. Most (> 99.9%) of the primordial follicles present at birth never ovulate, but undergo atresia, which can occur at any stage of follicular development. In humans, about 6 million oogonia are formed during fetal development; at birth the ovaries contain approximately 2 million oocytes incorporated into primordial follicles; by the time of puberty about 400,000 follicles remain; only about 400 follicles are destined to ovulate. Due to the ongoing significant loss of follicles by atresia, as a woman ages, her primordial follicle reserve dwindles and is ultimately depleted, resulting in ovarian failure (menopause), which is associated with the cessation of ovarian cyclicity. The average age of menopause in the United States is 51, and this is a direct consequence of depletion of the follicular reserve.

The stages of follicular development involve a continuum of events, each providing further maturation. Upon receipt of an as yet unknown signal for development, the primordial follicle is activated and becomes a primary follicle. As the follicle develops, there is proliferation of the granulosa cells surrounding the oocyte, and acquisition of a layer of theca interna cells surrounding the granulosa layer. Follicles progress from the primary stage to the secondary stage, when granulosa cells proliferate to form multiple layers around the oocyte. Collectively, primordial, primary, and secondary follicles are classified as preantral. When the follicle develops sufficiently, an antrum (fluid-filled space) develops within the granulosa cell layer. The antral follicle continues to grow, and at its most mature stage prior to ovulation is known as a Graafian (preovulatory) follicle. Following ovulation, the cells remaining, which formed the structure of the follicle, infiltrate and differentiate (luteinize) to form a solid gland, the corpus luteum. Resumption of meiosis in the oocyte occurs only at the time of impending ovulation.

Sites of Disruption of Ovarian Function

A reproductive toxicant can have detrimental effects on ovarian function at several levels. It can cause a direct ovarian effect, or an indirect effect on the hypothalamus and/or pituitary. Acting through a direct effect, compromised ovarian function would result in reduced ovarian steroid production. A decrease in circulating steroids would relieve negative feedback on the hypothalamus and pituitary. As a result, gonadotropin (LH and FSH) levels would increase. Conversely, if the effect occurs at the hypothalamic-pituitary level, reductions in gonadotropin secretion would precede ovarian steroid changes. With respect to environmental toxicants that directly target the ovary, those that extensively destroy primordial and primary follicles (small preantral) can cause premature ovarian failure (early menopause in women). Once destroyed, primordial follicles cannot be replaced. Furthermore, such destruction will have a delayed effect on ovarian cyclicity that is undetected until there are no follicles left to be recruited for development. Alternatively, an environmental toxicant that selectively damages large growing or antral follicles can cause a reversible disruption of cyclicity by impairing ovarian steroid production and ovulation. This effect can be reversible because remaining primordial follicles will ultimately be recruited for development if exposure to the toxicant ceases.

Consequences of Ovarian Toxicity

The level and duration of exposure to an environmental toxicant can determine the effects on reproduction. Individuals are only rarely exposed acutely to high levels of toxicants. These can usually be readily identified. However, the possible effects of chronic, low dose exposures are of particular concern, because they are more likely to occur and are more difficult to identify. These types of exposures may cause reproductive or fertility problems that go unrecognized for years due to the potential for cumulative damage. Manifestations of such injury could include infertility, early menopause, or eventual development of ovarian cancer.

Another factor involving the overall effects of reproductive toxicants is the lifecycle stage at which exposure occurs. Temporary infertility can occur in an adult cyclic woman following damage to the ovaries, whereas exposure during childhood might induce sterility by chemical-induced destruction of germ cells. Furthermore, exposures *in utero* may cause improper development of ovarian follicles or permanent alterations in the reproductive tract.

Metals: Ovarian Effects

Cadmium (Cd)

Cadmium is the most extensively reported metal with respect to effects on ovarian function. Sources of exposure of humans to cadmium in the environment are widespread. One particular source is cigarette smoke (Zenzes et al., 1995; Piasek et al., 2001; Younglai et al., 2002). However, cadmium exposure in the environment also occurs via food, modern industrial processes, waste disposal, and terrestrial and aquatic ecosystems (Nath et al., 1984; Piasek and Laskey, 1999).

The result of exposure in humans is often estimated by measurement of cadmium in ovarian follicular fluid of women undergoing assisted reproductive technologies to treat infertility. In one study, cadmium was higher in follicular fluid in women who were smokers vs. nonsmokers (Zenzes et al., 1995). It was suggested that this accumulation might compromise the quality of oocytes, becoming a risk factor for fertility. Interestingly, results of two studies in which fertility was evaluated in women with high vs. low levels of cadmium in follicular fluid did not support this conclusion (Drbohlav et al., 1998; Younglai et al., 2002). In those studies, there was no difference in conception rates between the two groups. Thus, these studies supported the theory that fertility is not adversely affected by cadmium accumulated in the ovarian follicle.

Animal studies have also demonstrated that the ovary is a site of accumulation of cadmium following experimentally induced exposure. Following injection, significant levels of cadmium were measured in ovarian follicle walls of freshwater painted turtles and laying hens (Rie et al., 2001; Sato et al., 1996). This was associated with increases in expression of metal-binding proteins thought to sequester the divalent cation at that site. Administration of relatively low doses of cadmium to mice or rats (0.25 to 1 mg/kg) resulted in accumulation in a number of tissues, including the ovary (Massanyi et al., 1999; Varga et al., 1991). This level of exposure was insufficient to produce alterations in ovarian production of progesterone or 17β-estradiol (Varga et al., 1991). Conversely, cadmium chloride administered to rats at higher levels (3.5 or 7.0 mg/kg) during pseudopregnancy was rapidly incorporated into corpora lutea, and the higher dose caused a significant decrease in circulating progesterone levels when given during mid-pseudopregnancy (Paksy et al., 1990a). The route of administration of cadmium appears to affect its tissue distribution. Levels of cadmium in ovaries of rabbits given 1.5 mg/kg intraperatoneally (i.p.) were 174 times higher than those in control animals, whereas oral exposure to cadmium resulted in ovarian levels that were only 16 times those in controls (Massanyi et al., 1995).

Several lines of evidence support the theory that ovarian accumulation of cadmium causes direct adverse effects. Morphological evaluations have demonstrated hemorrhagic necrosis in hamster ovaries following injection of cadmium chloride at about 5 mg/kg (Saksena and Salmonsen, 1983; Rehm and Waalkes, 1988; Waalkes and Rehm, 1998). However, this toxicity did not result in tumor development (Waalkes and Rehm, 1998). Mice and rats were less susceptible than hamsters to cadmium in development of these lesions (Rehm and Waalkes, 1988). However, in mice, atretic follicle and degenerating corpora lutea were also seen following exposure to cadmium (Godowicz and Pawlus, 1985). An impact of cadmium on oocyte development, as well as atresia, has also been reported in Xenopus laevis (Lienesch et al., 2000; Fort et al., 2001), with increased necrosis being observed at levels above 5 mg/kg. This appears to be a widely conserved effect, as deleterious effects of cadmium on oocytes in red swamp crayfish and earthworms have also been reported (Reddy et al., 1997; Siekierska and Urbanska-Jasik, 2002).

Cadmium exposure has also been reported to impact ovarian steroidogenesis. Acute exposure of rats to cadmium (up to 5 mg/kg) resulted in decreases in circulating progesterone and 17β-estradiol, and this was cycle stage-dependent (Piasek et al., 1996). Additionally, the rise in progesterone production associated with mid-pregnancy in rats was prevented by up to 10 mg/kg cadmium exposure on day 1 after conception (Paksy et al., 1992a). Decreases in circulating steroid levels as a result of cadmium exposure might be caused by direct ovarian effects, or decreased gonadotropin secretion from the anterior pituitary. Thus, those measurements alone cannot identify the physiological site targeted.

Support that the effects of cadmium on steroidogenesis result from direct ovarian targeting is provided by *in vitro* studies. Whole ovarian cell cultures from rats demonstrated particular susceptibility to *in vitro* cadmium exposure when tissues were collected during proestrus or early pregnancy. During those stages, progesterone and testosterone (Piasek et al., 2002; Piasek and Laskey, 1999) as well as 17β-estradiol (Piasek and Laskey, 1994) decreased when cadmium was added to the medium. The effect was observed whether or not human chorionic gonadotropin was included in the incubation (Piasek and Laskey, 1999). In another approach, rat ovarian granulosa and luteal cells were separated for determination of the *in vitro* impact of cadmium on steroidogenesis (Paksy et al., 1992b). FSH-stimulated progesterone and estrogen secretion were inhibited in granulosa cells, while LH-stimulated progesterone production was inhibited in luteal cells. Cadmium had no effect on basal progesterone production in either cell type.

Evidence for a hypothalamic-pituitary site of cadmium targeting has also been provided in studies that measured circulating gonadotropin levels or evaluated the effect of exposure on ovulation. Effects of cadmium at those sites would likely manifest as decreases in FSH and LH secretion, and an ability to stimulate ovulation with exogenous gonadotropins. In golden hamsters, cadmium chloride (50 to 10 mg/kg) was given close to the time of the

LH surge that triggers ovulation (Saksena and Salmonsen, 1983). This resulted in failure of ovulation and decreases in circulating progesterone levels. Both of these events would not likely be observed unless the LH surge had been retarded. Similar findings were observed in rats in response to acute or long-term exposure to cadmium (Paksy et al., 1989; 1990b). Acute exposure to cadmium resulted in reduced serum levels of LH and FSH on the expected day of estrus. Additionally, long-term exposure resulted in anovulatory cycles. Anovulation was confirmed by reduced oviductal retrieval of ovulated oocytes. The most convincing evidence for a hypotha-lamic-pituitary site of cadmium effects was provided in a study in which rats were evaluated during proestrus following exposure to cadmium during diestrus (Varga and Paksy, 1991). At evaluation, there was a reduced pituitary content of LH, however, stimulation of ovulation could be achieved by injection of exposed rats with luteinizing hormone releasing hormone (LHRH, hypothalamic LH releasing factor). This demonstrated that the likely effect of cadmium in this study was a reduction in endogenous LHRH because an exogenous source could by-pass the impairment of ovulation. One study in fiddler crabs supported a neuroendocrine site of cadmium targeting (Rodr et al., 2000). The observations reported in that study dem-onstrated a requirement of neuronal gonad-stimulating hormone (GSH) to demonstrate cadmium-induced retardation of ovarian oocyte development. Taken together, the evidence for cadmium-induced effects on ovarian func-tion support targeting of both ovarian and neuroendocrine regulation.

Little is known about the molecular mechanisms of cadmium-induced ovarian toxicity. However, there have been several interesting studies dem-onstrating a direct competition of cadmium for endogenous metals known to be required to facilitate normal cellular functions. At the molecular level, cadmium interferes with the utilization of essential metals such as calcium, zinc, selenium, chromium, and iron (Nath et al., 1984). Deficiencies in these essential metals have been shown to exaggerate cadmium toxicity. Cadmium transport across the intestinal and renal brush border membrane vesicles is carrier mediated and competes with zinc and calcium (Nath et al., 1984). In placentas of smoking women there was an association between increased cadmium and decreased iron concentrations (Piasek et al., 2001). A metal binding protein was isolated from ovaries of dab fish (Limanda limanda L.) caught in the North Sea (Kammann et al., 1996). Zinc binding to the protein was displaced by cadmium and dose-dependent amounts of cadmium were bound to the protein after exposure of dab to cadmium. Additionally, pre-treatment of Syrian hamsters with zinc acetate protected against cad-mium-induced ovarian lesions (Rehm and Waalkes, 1988).

In summary, although there is no compelling evidence to link cadmium toxicity with impaired fertility in humans, animal studies have demonstrated there is potential for adverse effects on ovarian function. There is evidence for ovarian accumulation, follicle loss by necrosis or atresia, and reductions in steroid production. Collectively these events, if produced in humans, would potentially impact conception rates by reducing ovulation and impairing progesterone-related events such as implantation and maintenance of pregnancy.

Lead (Pb)

Exposure to lead is widespread in the environment. The greatest source of exposure is in commercial products such as paint, gasoline, printing material, and acid batteries (Dearth et al., 2002). High levels of exposure also occur in many industries, such as mining and refining plants (Tapeau et al., 2001). A major concern related to exposure to lead is its impact on fertility in women, and on fetal development and prepubertal growth retardation; thus lead exposure/toxicity is a leading environmental health issue for children and women of child bearing age (Tapeau et al., 2003; Dearth et al., 2002). Monitoring of blood levels has generally been undertaken to provide evidence of exposure levels. A cutoff level of 10 micrograms/dL has been set as a predicted risk accumulation of circulating lead levels (Crocetti et al., 1990). Based on that, it was estimated that 4.4 million U.S. women of childbearing age were at risk in the late 1980s, with 10% of those women becoming pregnant each year. Because there has been a heightened awareness of lead in the environment, exposure potentials have been declining.

Related to ovarian effects caused by lead, there are limited reports of its accumulation in follicular fluid in women. One study compared lead content in follicular fluid with *in vitro* effects of lead on granulosa cells collected from women undergoing *in vitro* fertilization and embryo transfer (Paksy et al., 2001). Although levels of lead in the nanomolar range were detectable in follicular fluid, millimolar concentrations were required to decrease progesterone secretion in cultured granulosa cells collected from those follicles. The authors concluded that lead levels measured in ovarian follicular fluid are unlikely to pose a hazard with respect to progesterone secretion by the ovary. However, lead effects reported in human pregnancies relate mostly to miscarriages, premature delivery, and infant mortality (Taupeau et al., 2001). These outcomes suggest direct toxicities on placental or fetal development rather that ovarian effects. Accumulation in tissue was monitored in mice exposed to lead (Taupeau et al., 2001). The ovaries contained significantly increased concentrations relative to controls, and this was about five times greater in mice that were acutely, as compared with chronically, exposed to the same concentration of lead (10 mg/kg, 15 days vs. 15 weeks, respectively). Lead exposure in fathead minnows was evaluated in a comparative study (Weber, 1993). High lead accumulations occurred in ovaries of fish maintained in water with 0.5 ppm lead for 4 weeks. This was associated with decreased numbers of eggs oviposited, increased interspawn periods, and suppressed embryo development. Thus, evidence for ovarian effects in fish has been reported.

Morphological evidence of direct ovarian effects of lead exposure in mice has been described. Daily treatment of adult females with lead nitrate

(10 mg/kg) for 15 days or 15 weeks, or with lead acetate (2 to 8 mg/kg), resulted in destruction of small preantral follicles and increased atresia in ovaries of exposed mice (Taupeau et al., 2001; Junaid et al., 1997). Furthermore, transplacental exposure to lead resulted in a reduced number of primordial follicles in female pups (Wide, 1985). Other animal studies have supported functional impairment of steroidogenesis resulting from lead exposure. Female rhesus monkeys were exposed to lead acetate in drinking water for 33 months (Franks et al., 1989). Relative to controls, they exhibited longer and more variable menstrual cycles and shorter menstrual flow. Circulating levels of progesterone were reduced, although ovulatory cycles were not different. The authors concluded that lead given in this study did not prevent ovulation, but luteal function was suppressed.

Transplacental effects of lead have shown an impact on female F1 offspring. *In utero* exposure in Sprague-Dawley rats resulted in decreased ovarian conversion of progestereone to androstenedione in prepubertal and pubertal females (Wiebe et al., 1988). It was concluded that prenatal exposure to lead significantly altered subsequent ovarian steroid production and gonadotropin binding. In other transplacental studies in rats, it was concluded that 17β-estradiol production during puberty was impaired following *in utero* exposure to lead (Ronis et al., 1998; Srivastava et al., 2003). Whether these effects were directly at the ovarian level could not be predicted from these findings. One study in humans provided evidence for direct ovarian impairment of steroid production. Expression of cytochrome p450 aromatase that converts androgen to estrogen, and the estrogen receptor-beta mRNA and protein was significantly reduced by *in vitro* lead exposure of granulosa cells collected from women undergoing *in vitro* fertilization (Tapeau et al., 2003). As a different mechanistic explanation of the effects of lead on steroidogenesis, reduced expression of ovarian steroidogenic acute regulatory protein (StAR) was seen (Srivastava et al., 2003). Because pregnant mares' serum (exogenous gonadotropin) could restore StAR expression and 17β-estradiol production, it was concluded that this effect was the result of reduced pituitary support of ovarian function. This finding provides support for a hypothalamic-pituitary rather than direct ovarian site of action.

In further support of a neuroendocrine targeting by lead, several studies have focused on pubertal development in rats. *In utero* exposure to lead resulted in delayed timing of puberty in female offspring, which was associated with suppressed serum levels of LH and 17β-estradiol (Dearth et al., 2002). In another study, prepubertal exposure to lead resulted in delayed vaginal opening and disrupted estrous cyclicity (Ronis et al., 1996). In females exposed *in utero*, these effects were also associated with decreased circulating 17β-estardiol and LH, and reduced pituitary content of LH protein and LH beta mRNA. These results prompted the authors to suggest a dual site of lead action at the hypothalamic-pituitary unit as well as ovarian steroidogenesis.

In summary, studies reporting the effects of lead exposure on female reproductive function suggest a direct targeting of small preantral follicles, as well as impaired steroidogenesis by direct ovarian and indirect hypotha-lamic-pituitary targeting. Even though there is no convincing evidence for an oocyte-direct effect of lead in humans, reduced fertility in exposed women might result from reduced 17β-estradiol to support follicular development, and reduced progesterone to provide support for implantation and mainte-nance of pregnancy. The data do not support reduced ovulation (amenor-rhea). Rather, the predicted effects in women would be delayed puberty, reductions in pregnancy establishment, and early menopause.

Mercury (Hg)

Agriculture, consumption of fossil fuels, and industry are the main sources of pollution by mercury in the environment (Leonard et al., 1983). These are largely in the form of metallic mercury as inorganic or organic compounds. Additionally, ingestion of organomercurials from eating fish can be absorbed via the gastrointestinal tract. Elemental mercury is a liquid, and is found in thermometers, fluorescent lightbulbs, barometers, blood pressure instru-ments, and mercury switches in children's shoes that light up (Davis et al., 2001). Furthermore, dentists, their assistants, and patients are exposed to elemental mercury vapor in the preparation of dental silver amalgams (reviewed by Schuurs, 1999). Exposure to elemental mercury is largely via inhalation. In an Italian mercury vapor lamp factory, women were chroni-cally exposed to mercury inhalation (DeRosis et al., 1985). Increased irregu-larities in the menstrual cycle were reported in the subset of women exposed to high levels of mercury in conjunction with increased stress, relative to the control group. Thus, there is evidence for ovarian effects caused by lead in women. All chemical forms of mercury given to animals have been shown to cause reproductive problems including infertility, disturbances in the menstrual cycle, and inhibition of ovulation—all potentially arising from ovarian targeting (Schuurs, 1999). Alternatively, exposure to mercury may cause hormonal effects by targeting a hypothalamic-pituitary site. Mercury has been shown to accumulate in the pituitary (Lamperti and Printz, 1974; Erfurth et al., 1990). However, in a study in men there was no evidence of an association between exposure to mercury and decreased serum concen-trations of testosterone (Erufurth et al., 1990). This predicts that there would be a similar lack of association with ovarian steroids in women. Although several studies conducted mostly in Europe have reported menstrual cycle abnormalities in women occupationally exposed to mercury vapor, there is a lack of convincing evidence in the literature to suggest that exposure to mercury causes ovarian or hypothalamic-pituitary effects (Davis et al., 2001). This points up a need for more studies to understand how mercury may affect the reproductive system in women.

Ovarian accumulation of mercury has been reported in animal studies. Following exposure to mercury chloride, mercury accumulation in various hamster tissues was greatest in the kidney, followed by the liver, anterior pituitary, ovary, blood, uterus, hypothalamus, and the cerebral cortex

(Lamperti and Printz, 1974). Within the ovary, mercury accumulation was observed in corpora lutea in mice and hamsters (Khayat and Dencker, 1982; Lamperti and Printz, 1974). Likewise, in a study in rats exposed to mercuric chloride, deposits of mercury were highly concentrated within macrophages of atretic young and mature ovarian follicles and corpora lutea, as well as within granulosa cells of atretic follicles and lutein cells of freshly formed corpora lutea (Stadnicka, 1980). Mercury also accumulated in ovaries of rats exposed to elemental mercury by inhalation (Davis et al., 2001). In another report, high levels of ovarian accumulation of mercury were observed in prespawning fish (Labeo rohita) exposed to sublethal levels of mercuric chloride (Aditya et al., 2002). In a study of female golden hamsters, a comparison of ovarian mercury accumulation was made between treatment with methylmercuric chloride and mercuric chloride (Watanabe et al., 1982). There was a greater concentration of ovarian mercury in animals treated with methylmercuric chloride when compared with those treated with mercuric chloride. Conversely, serum concentrations were higher in animals treated with mercuric chloride. Interestingly, mercuric chloride treatment inhibited ovulation, whereas methymercuric chloride treatment did not. The authors concluded that either mercuric chloride is more directly toxic to the ovary than methylmercury, or higher circulating levels of mercuric chloride are responsible for targeting the hypothalamic-pituitary unit thereby causing disruptions in ovulation. Thus, as well as the ovary, the pituitary could be a site of mercury action.

Animal studies have provided evidence for both sites as target(s) of mercury-induced effects. Morphological evaluation of ovaries from exposed animals demonstrated an immature appearance of corpora lutea in rats exposed by inhalation (4 mg/m³ for 11 days, Davis et al., 2001). Circulating 17β-estradiol and progesterone levels were significantly reduced. However, this was thought to be a result of generalized toxicity because there was also weight loss in animals exposed to that dose of mercury. In a group exposed to a lower dose (1 or 2 mg/m³ for 8 days), there was no effect on serum steroid levels. The observation in rats exposed to the higher dose of mercury was similar to that in female hamsters in which ovaries from animals treated with mercury chloride (1 mg/day for 4 days) showed retarded follicular development and morphologically prolonged corpora lutea (Lamperti and Printz, 1973). These effects were accompanied by reductions in circulating progesterone. However, in that study, body weights were not reported. Therefore, it is difficult to assess the specificity of these reported effects. If they were specific, these collective observations suggest a lack of appropriate trophic support for follicular development and luteinization from pituitary hormones.

Mercuric chloride exposure has been shown to disrupt the estrous cycle in rats and cause anovulation in hamsters (Stadnicka, 1980; Lamperti and Printz, 1974). In a study in which rats were exposed to elemental mercury by inhalation, estrous cycles were prolonged (Davis et al., 2001). There are several reports of the effects of mercury exposure on oocyte morphology.

There was a decrease in oocytes ovulated and an increase in their degeneration in golden hamsters treated with mercuric chloride (Watanabe et al. 1982), although no chromosomal aberrations in the oocytes were observed. A similar finding was reported in mice treated with mercury chloride (Shen et al., 2000).

One source of concern with respect to exposure to mercury in humans is eating fish that have been living in contaminated waters. However, fish may also be at risk for ovarian effects of mercury. Exposure of Channa punctatus and Labeo rohita to mercury resulted in decreases in more developed oocytes (Dey and Bhattacharya, 1989; Aditya et al., 2002). There was a decrease in ovarian content of protein, lipids, and cholesterol in Notopterus notopterus exposed to mercuric chloride (Sindhe et al., 2002). In Channa punctatus, ovarian alkaline phosphatase, glucose-6-phosphatase, and lipase activities were reduced following exposure to mercuric chloride (Sastry and Agrawal, 1979). Ovarian maturation following mercury exposure in red swamp crayfish was inhibited (Reddy et al., 1997). This effect was attributed to targeting of neuroendocrine regulation. Thus, in addition to fish as a vehicle for human exposure to mercury, the impact of contaminated waters on their reproductive capacity should also be of concern.

Collectively, in humans the evidence for mercury-induced effects on ovarian function is not strong. Furthermore, the results of animal studies have not provided sufficient mechanistic information to determine whether compromised ovarian function is the result of direct targeting of the ovary or indirect targeting of the hypothalamic-pituitary unit. More research is required to provide a clearer prediction as to the actual risk for ovarian impairment from exposure to mercury. From the observed effects in animal studies, the predicted impact in women would be disruptions in the menstrual cycle.

Chromium (Cr)

Chromium is an essential element in the human body required for proper carbohydrate, protein, and fat metabolism. Increasing chromium in the environment is a result of chromium-based industries (Junaid et al., 1996). Furthermore, workers in these industries can be exposed to concentrations two orders of magnitude higher than the general population. Concerns related to reproduction in humans largely center around fetotoxicity during pregnancy. Abnormal menses has been reported as a possible effect of chromium on ovarian function in women (Murthy et al. 1996). There have been few animal studies related to reproductive effects of chromium exposure. Additionally, there is little evidence to demonstrate ovarian targeting by this metal.

One study evaluating transplacental exposure in mice provided evidence for ovarian effects in the mothers (Junaid et al., 1996). Groups of female mice were exposed to increasing concentrations of potassium dichromate in the drinking water for 20 days prior to mating. At the time of parturition, there was 100% fetal loss in mothers treated with the high dose of chromium (750 ppm). The fetal loss at that dose was accompanied by a 45% reduction in numbers of corpora lutea in the maternal ovaries, which reflects a direct reduction in ovulation numbers. The absence of fetuses was mainly due to fetal toxicity. However, these data also provide evidence that chromium exposure caused a reduction in ovulation rates in the mothers, thus a targeting of ovarian function. In another study, evaluation of ovarian effects of chromium on nonpregnant female mice was made (Murthy et al., 1996). Following 20 days of exposure to increasing concentrations of potassium dichromate in the drinking water, ovarian preantral follicles were counted. Chromium caused a dose-dependent reduction in all stages of preantral follicles (primordial, primary, and growing) and number of oocytes ovulated, as well as a lengthening of the estrous cycle. There was also morphological evidence of follicular disruptions. These data support the observation related to fewer corpora lutea in the pregnant mouse study. Furthermore, the data suggest that high levels of chromium might disrupt menstrual cyclicity, ovulation and fertility in exposed women, as well as potentially cause early menopause.

Arsenic (As)

Arsenic has been increasing as a major pollutant in drinking water during the last decade in India, Bangladesh, Chile, Thailand, Taiwan, China, Inner Mongolia, Mexico, Argentina, Finland, and Hungary (Chattopadhyay et al., 2003). Additionally, there is growing concern related to arsenic contamination in the drinking water in the United States (Chappell et al., 1997). Reports from the Ukraine, Taiwan, and Bangladesh have linked arsenic-contaminated drinking water to reproductive disorders in women (Chattopadhyay et al., 2003). Few animal studies have provided evidence for arsenic-induced effects on ovarian function. Therefore, further evaluation of the impact of arsenic exposure on reproductive function is needed.

Arsenic accumulation in the ovary was shown in rats exposed to sodium arsenite in the drinking water for 28 days (Chattopadhyay et al., 1999; Chattopadhyay et al., 2003). There were greater levels of arsenic in the ovary, when compared to blood in both control and treated animals. Relative to control animals, ovarian content of arsenic was over twice as high in treated animals. Interestingly, sodium selenite given in combination with arsenic prevented the ovarian accumulation of arsenic (Chattopadyay et al., 2003). This was suggested to be due to trapping of arsenic by selenium, thereby resulting in low tissue uptake.

Several studies have reported effects of arsenic exposure on ovarian function in rats. There was a significant reduction in circulating LH, FSH, and 17β-estradiol in rats given 0.4ppm arsenic for 28 days. There was also a decrease in ovarian activities of the steroidogenic enzymes, 5-3 beta- and 17 beta- hydroxysteroid dehydrogenase, supporting the decrease in steroid production. Arsenic-treated rats also demonstrated reduced ovarian weights and prolonged estrous cyclicity. These effects were not observed after 16 days of treatment. It was concluded that duration of arsenic treatment is the critical factor for its adverse effect on ovarian activities. In subsequent studies, these same effects of arsenic exposure in rats were observed, and coadministration with L-ascorbate (vitamin c) or sodium selenite prevented these adverse ovarian effects (Chattopadhyay et al., 2001; Chattopadhyay et al., 2003). It was concluded that arsenic-induced reproductive toxicity may be due to the induction of oxidative stress or free radical generation, and these agents may serve an antioxidant function. The results of these studies suggest that arsenic may have direct ovarian effects; however in the face of reduced gonadotropin levels, a hypothalamic-pituitary site of targeting is also likely.

Transplacental exposure to high levels of arsenic was also observed to induce ovarian epithelial tumors in mice (Waalkes et al., 2003a; Waalkes et al., 2003b). Mice were exposed during gestational days 8 to 18 when mothers were given high concentrations of sodium arsenite (42.5 and 85 ppm) in the drinking water.

When arsenic-exposed mice were given the tumor promoter, 12-O-tetradecanoyl phorbol-13-acetate (TPA) at 4 to 25 weeks of age, they developed hepatic, lung, and ovarian epithelial tumors Waalkes et al., 2003b). Interestingly, the ovarian tumors were accompanied by hyperplasia of the uterus and oviduct, and these effects developed independently of TPA treatment. These findings establish that gestation is a period of high sensitivity to arsenic-induced carcinogenesis, and that the female reproductive system is highly susceptible.

Taken together, the studies reported concerning arsenic exposure suggest that it would be capable of disrupting ovarian cyclicity, likely via a hypothalamic-pituitary site of action. Additionally, *in utero* exposure to very high concentrations may result in development of reproductive cancers in women.

Summary and Future Directions

In summary, animal studies have provided evidence that several metals found in the environment have the potential to affect ovarian function either directly or indirectly via hypothalamic-pituitary targeting. There is evidence for ovarian accumulation of these metals, and in some cases they have been detected in follicular fluid in women. However, there is no concrete evidence for adverse ovarian effects caused by any of these metals in women. Table 7.1 summarizes the evidence for sites of targeting of ovarian function and predictions of the nature of adverse effects in women that would be expected in each case. Future studies should involve more mechanistic animal studies, as well as detailed epidemiological investigations to determine whether environmental exposures to these metals actually pose a risk to ovarian function in women.

Table 7.1 Sites of Targeting and Predicted Effects of Cd, Pb, Hg, Cr, and As on Reproductive Function in Women

Metal	Cd	Pb	Hg	Cr	As
Ovarian Accumulation:	+	+	+	?	+
Target:					
Ovary	+	+	?	?	+
Hypothalamus-Pituitary	+	+	?	?	+
Animal Effects:					
Oocyte loss	+	+	+	–	–
Follicular atresia	+	+	+	–	–
↓ Steroidogenesis	+	+	+	–	–
Ovarian cancers	–	–	–	–	+
Predicted Outcomein Women:					
Disrupted menses	+	–	+	+	+
Infertility	+	+	–	+	–
Early menopause	–	+	–	+	–

+ = observed effects support

– = observed effects do not support

? = insufficient information available

References

Aditya, A.K., Chattopadhyay, S., and Mitra, S. (2002) Effect of mercury and methyl parathion on the ovaries of Labeo rohita (Ham), *J. Environ. Biol.*, 23, 61–64.

Chappell, W.R., Beck, B.D., Brown, K.G., Chaney, R., Cothern, R., Cothern, C.R., Irgolic, K.J., North, D.W., Thornton, I., and Tsongas, T.A. (1997) Inorganic arsenic: a need and an opportunity to improve risk assessment, *Environ. Health Perspect.*, 105, 1060–67.

Chattopadhyay, S., Ghosh, S., Chaki, S., Debnath, J., and Ghosh, D. (1999) Effect of sodium arsenite on plasma levels of gonadotrophins and ovarian steroidogenesis in mature albino rats: duration-dependent response, *J. Toxicol. Sci.*, 24, 425–31.

Chattopadhyay, S., Ghosh, S., Debnath, J., and Ghosh, D. (2001) Protection of sodium arsenite-induced ovarian toxicity by coadministration of L-ascorbate (vitamin C) in mature wistar strain rat, *Arch. Envrion. Contam. Toxicol.*, 41, 83–9.

Chattopadhyay, S., Pal Ghosh, S., Ghosh, D., and Debnath, J. (2003) Effect of dietary co-administration of sodium selenite on sodium arsenite-induced ovarian and uterine disorders in mature albino rats, *Toxicol. Sci.*, 75, 412–22.

Crocetti, A.F., Mushak, P., and Schwartz, J. (1990) Determination of numbers of lead-exposed women of childbearing age and pregnant women: an integrated summary of a report to the U.S. Congress on childhood lead poisoning, *Environ. Health Perspect.*, 89, 121–24.

Davis, B.J., Price, H.C., O'Connor, R.W., Fernando, R., Rowland, A.S., and Morgan, D.L. (2001) Mercury vapor and female reproductive toxicity, *Toxicol. Sci.*, 59, 291–96.

Dearth, R.K., Hiney, J.K., Srivastava, V., Burdick, S.B., Bratton, G.R., and Dees, W.L. (2002) Effects of lead (Pb) exposure during gestation and lactation on female pubertal development in the rat, *Reprod. Toxicol.*, 16, 343–52.

De Rosis, F., Anastasio, S.P., Selvaggi, L, Beltrame, A., and Moriani, G. (1985) Female reproductive health in two lamp factories: effects of exposure to inorganic mercury vapour and stress factors, *Br. Ind. Med.*, 42, 488–94.

Dey, S. and Bhattacharya, S. (1989) Ovarian damage to Channa punctatus after chronic exposure to low concentrations of Elsan, mercury, and ammonia, *Ecotoxicol. Environ. Saf.*, 17, 247–57.

Drbohlav, P., Bencko, V., Masata, J., Bendl, J., Rezacova, J., Zouhar, T., Cerny, V., and Halkova, E. (1998) Detection of cadmium and zinc in the blood and follicular fluid in women in the IVF and ET program, *Ceska Gynekol.*, 63, 292–300.

Erfurth, E.M., Schutz, A., Nilsson, A., Barregard, L., and Skerfving, S. (1990) Normal pituitary hormone response to thyrotrophin and gonadotrophin releasing hormones in subjects exposed to elemental mercury vapour, *Br. J. Ind. Med.*, 47, 639–44.

Fort, D.J., Stover, E.L., Bantle, J.A., Dumont, J.N., and Finch, R.A. (2001) Evaluation of a reproductive toxicity assay using *Xenopus laevis*: boric acid, cadmium and ethylene glycol monomethyl ether, *J. Appl. Toxicol.*, 21, 41–52.

Franks, P.A., Laughlin, N.K., Dierschke, D.J., Bowman, R.E., and Meller, P.A. (1989) Effects of lead on luteal function in rhesus monkeys, *Biol. Reprod.*, 41, 1055–62.

Godowicz, B. and Pawlus, M. (1985) Effect of cadmium chloride on the ovulation and structure of ovary in the inbred KP and CBA mice strains, *Folia Histochem. Cytobiol.*, 23, 209–15.

Hirshfield, A.N. (1991) Development of follicles in the mammalian ovary, *Int. Rev. Cytol.*, 124, 43–101.

Hoyer, P.B. and Devine, P.J. (2001) Endocrinology and toxicology: the female reproductive system, in M. Derelanko and M. Hollinger (eds.) *Handbook of Toxicology*, 2nd ed., CRC Press, Boca Raton, FL, 573–96.

Junaid, M., Chowdhuri, D.K., Narayan, R., Shanker, R., and Saxena, D.K. (1997) Lead-induced changes in ovarian follicular development and maturation in mice, *J. Toxicol. Environ. Health*, 50, 31–40.

Junaid, M., Murthy, R.C., and Saxena, D.K. (1996) Embryo- and fetotoxicity of chromium in pregestationally exposed mice, *Bull. Environ. Contam. Toxicol.*, 57, 327–34.

Kammann, U., Friedrich, M., and Steinhart, H. (1996) Isolation of a metal-binding protein from ovaries of dab (Limanda limanda L.) distinct from metallothionein: effect of cadmium exposure, *Ecotoxicol. Environ. Saf.*, 33, 281–86.

Khayat, A. and Dencker, L. (1982) Fetal uptake and distribution of metallic mercury vapor in the mouse: influence of ethanol and aminotriazole, *Int. J. Biol. Res. Pregnancy*, 3, 38–46.

Lamperti, A.A. and Printz, R.H. (1973) Effects of mercuric chloride on the reproductive cycle of the female hamster, *Biol. Reprod.*, 8, 378–87.

Lamperti, A.A. and Printz, R.H. (1974) Localization, accumulation, and toxic effects of mercuric chloride on the reproductive axis of the female hamster, *Biol. Reprod.*, 11, 180–86.

Leonard, A., Jacquet, P., and Lauwerys, R.R. (1983) Mutagenicity and teratogenicity of mercury compounds, *Mut. Res.*, 114, 1–18.

Lienesch, L.A., Dumont, J.N., and Bantle, J.A. (2000) The effect of cadmium on oogenesis in Xenopus laevis, *Chemosphere*, 41, 1651–58.

Massanyi, P., Bardos, L., Oppel, K., Hluchy, S., Kovacik, J., Csicsai, G., and Toman, R. (1999) Distribution of cadmium in selected organs of mice: effects of cadmium on organ contents of retinoids and beta-carotene, *Acta Physiol. Hung.*, 86, 99–104.

Massanyi, P., Toman, R., Valent, M., and Cupka, P. (1995) Evaluation of selected parameters of a metabolic profile and levels of cadmium in reproductive organs of rabbits after an experimental administration, *Acta Physiol. Hung.*, 83, 267–73.

Murthy, R.C., Junaid, M., and Saxena, D.K. (1996) Ovarian dysfunction in mice following chromium (VI) exposure, *Toxicol. Lett.*, 89, 147–54.

Nath, R., Prasad, R., Palinal, V.K., and Chopra, R.K. (1984) Molecular basis of cadmium toxicity, *Prog. Food Nutr. Sci.*, 8, 109–63.

Paksy, K., Gati, I., Naray, M., and Rajczy, K. (2001) Lead accumulation in human ovarian follicular fluid, and in vitro effect of lead on progesterone production by cultured human ovarian granulosa cells, *J. Toxiocol. Environ. Health A.*, 62, 359–66.

Paksy, K., Naray, M., Varga, B., Kiss, I., Folly, G., and Ungvary, G. (1990a) Uptake and distribution of Cd in the ovaries, adrenals, and pituitary in pseudopregnant rats: effect of Cd on progesterone serum levels, *Environ. Res.*, 51, 83–90.

Paksy, K., Varga, B., and Folly, G. (1990b) Long-term effects of a single cadmium chloride injection on the ovulation, ovarian progesterone and estradiol-17 beta secretion in rats, *Acta Physiol. Hung.*, 76, 245–52.

Paksy, K., Varga, B., and Lazar, P. (1992) Cadmium interferes with steroid biosynthesis in rat granulose and luteal cells in vitro, *Biometals*, 5, 245–50.

Paksy, K., Varga, B., Horvath, E., Tatrai, E., and Ungvary, G. (1989) Acute effects of cadmium on preovulatory serum FSH, LH and prolactin levels and on ovulation and ovarian hormone secretion in estrous rats, *Reprod. Toxicol.*, 3, 241–47.

Paksy, K., Varga, B., Naray, M., Olajos, F., and Folly, G. (1992a) Altered ovarian progesterone secretion induced by cadmium fails to interfere with embryo transport in the oviduct of the rat, *Reprod. Toxicol.*, 6, 77–83.

Paksy, K., Varga, B., Lazar, P. (1992b) Cadmium interferes with steroid biosynthesis in rat granulosa and luteal cells *in vitro*, *Biometals*, 5, 245–50.

Piasek, M. and Laskey, J.W. (1994) Acute cadmium exposure and ovarian steroidogenesis in cycling and pregnant rats, *Reprod. Toxicol.*, 8, 495–507.

Piasek, M. and Laskey, J.W. (1999) Effects of *in vitro* cadmium exposure on ovarian steroidogenesis in rats, *J. Appl. Toxicol.*, 19, 211–17.

Piasek, M., Blanusa, M., Kostial, K., and Laskey, J.W. (2001) Placental cadmium and progesterone concentrations in cigarette smokers, *Reprod. Toxicol.*, 15, 673–81.

Piasek, M., Laskey, J.W., Kostial, K., and Blanusa, M. (2002) Assessment of steroid disruption using cultures of whole ovary and/or placenta in rat and in human placental tissue, *Int. Arch. Occup. Environ. Health*, Suppl, S36–44.

Piasek, M., Schonwald, N., Blanusa, M., Kostial, K., and Laskey, J.W. (1996) Biomarkers of heavy metal reproductive effects and interaction with essential elements in experimental studies on female rats, *Arch. Hig. Rada Toksikol.*, 47, 245–59.

Reddy, P.S., Tuberty, S.R., and Fingerman, M. (1997) Effects of cadmium and mercury on ovarian maturation in the red swamp crayfish, Procambarus clarkii, *Ecotoxicol. Environ. Saf.*, 37, 62–65.

Rehm, S. and Waalkes, M.P. (1988) Cadmium-induced ovarian toxicity in hamsters, mice, and rats, *Fund. Appl. Toxicol.*, 10, 635–47.

Richards, J.S., Jahnsen, T., Hedin, L., Lifka, J., Ratoosh, S., Durica, J.M., and Goldring, N.B. (1987) Ovarian follicular development: from physiology to molecular biology, *Recent Progress in Hormone Research*, 43, 23–276.

Rie, M.T., Lendas, K.A., and Callard, I.P. (2001) Cadmium: tissue distribution and binding protein induction in the painted turtle, Chrysemys picta, *Comp. Biochem. Physiol. C Toxicol. Pharmacol.*, 130, 41–51.

Rodr, E.M., Greco, L.S., and Fingerman, M. (2000) Inhibition of ovarian growth by cadmium in the fiddler crab, Uca pugilator (Decapoda, ocypodidae), *Ecotoxicol. Environ. Saf.*, 46, 202–06.

Ronis, M.J., Badger, T.M., Shema, S.J., Roberson, P.K., and Shaikh, F. (1996) Reproductive toxicity and growth effects in rats exposed to lead at different periods during development, *Toxicol. Appl. Pharmacol.*, 136, 361–71.

Ronis, M.J., Badger, T.M., Shema, S.J., Roberson, P.K., Templer, L, Ringer, D., and Thomas, P.E. (1998) Endocrine mechanisms underlying the growth effects of developmental lead exposure in the rat, *J. Toxicol. Environ. Health A.*, 54, 101–20.

Saksena, S.K. and Salmonsen, R. (1983) Effects of cadmium chloride on ovulation and on induction of sterility in the female golden hamster, *Biol. Reprod.*, 29, 249–56.

Sastry, K.V. and Agrawal, M.K. (1979) Mercuric chloride induced enzymological changes in kidney and ovary of a teleost fish, Channa punctatus, *Bull. Environ. Contam. Toxicol.*, 22, 38–43.

Sato, S., Okabe, M., Kurasaki, M., and Kojima, Y. (1996) Metallothionein in the ovaries of laying hens exposed to cadmium, *Life Sci.*, 58, 1561–67.

Schuurs, A.H.B. (1999) Reproductive toxicity of occupational mercury. A review of the literature, *J. Dentistry*, 27, 249–56.

Shen, W., Chen, Y., Li, C., and Ji, Q. (2000) Effect of mercury chloride on the reproductive function and visceral organ of female mouse, *Wei Sheng Yan Jiu*, 29, 75–77.

Siekierska, E. and Urbanska-Jasik, D. (2002) Cadmium effect on the ovarian structure in earthworm Dendrobaena veneta (Rosa), *Environ. Pollut.*, 120, 289–97.

Sindhe, V.R., Veeresh, M.U., and Kulkarni, R.S. (2002) Ovarian changes in response to heavy metal exposure to the fish, Notopterus notopterus (Pallas), *J. Environ. Biol.*, 23, 137–41.

Srivastava, V., Dearth, R.K., Hiney, J.K., Ramirez, L.M., Bratton, G.R., and Dees, W.L. (2003) The effects of low-level Pb on steroidogenic acute regulatory protein (StAR) in the prepubertal rat ovary, *Toxicol. Sci.* n.p. (ahead of print).

Stadnicka, A. (1980) Localization of mercury in the rat ovary after oral administration of mercuric chloride, *Acta Histochem.*, 67, 227–33.

Taupeau, C., Poupon, J., Nome, F., and Lefevre, B. (2001) Lead accumulation in the mouse ovary after treatment-induced follicular atresia, *Reprod. Toxicol.*, 15, 385–91.

Taupeau, C., Poupon, J., Treton, D., Brosse, A., Richard, Y., and Machelon, V. (2003) Lead reduces messenger RNA and protein levels of cytochrome p450 aromatase and estrogen receptor beta in human ovarian granulosa cells, *Biol. Reprod.*, 68, 1982–88.

Varga, B. and Paksy, K. (1991) Toxic effects of cadmium on LHRH-induced LH release and ovulation in rats, *Reprod. Toxicol.*, 5, 199–203.

Varga, B., Paksy, K., and Naray, M. (1991) Distribution of cadmium in ovaries, adrenals and pituitary gland after chronic administration in rats, *Acta Physiol. Hung.*, 78, 221–26.

Waalkes, M.P. and Rehm, S. (1998) Lack of carcinogenicity of cadmium chloride in female Syrian hamsters, *Toxicol.*, 126, 173–78.

Waalkes, M.P., Ward, J.M., Liu, J., and Diwan, B.A. (2003a) Transplacental carcinogenicity of inorganic arsenic in the drinking water: induction of hepatic, ovarian, pulmonary, and adrenal tumors in mice, *Toxicol. Appl. Pharmacol.*, 186, 7–17.

Waalkes, M.P., Ward, J.M., and Diwan, B.A. (2003b) Induction of tumors of the liver, lung, ovary, and adrenal in adult mice after brief maternal gestational exposure to inorganic arsenic: promotional effects of postnatal phorbol ester exposure on hepatic and pulmonary, but not dermal cancers, *Carcinogenesis*, n.p. (ahead of print).

Watanabe, T., Shimada, T., and Endo, A. (1982) Effects of mercury compounds on ovulation and meiotic and mitotic chromosomes in female golden hamsters, *Teratology*, 25, 381–84.

Weber, D.N. (1993) Exposure to sublethal levels of waterborne lead alters reproductive behavior patterns in fathead minnows (Pimephales promelas), *Neurotoxicol.*, 14, 347–58.

Wide, M. (1985) Lead exposure on critical days of fetal life affects fertility in the female mouse, *Teratology*, 32, 375–80.

Wiebe, J.P., Barr, K.J., and Buckingham, K.D. (1988) Effect of prenatal and neonatal exposure to lead on gonadotropin receptors and steroidogenesis in rat ovaries, *J. Toxicol. Environ. Health*, 24, 461–76.

Younglai, E.V., Foster, W.G., Hughes, E.G., Trim, K., and Jarrell, J.F. (2002) Levels of environmental contaminants in human follicular fluid, serum and seminal plasma of couples undergoing in vitro fertilization, *Arch. Environ. Contam. Toxicol.*, 43, 121–26.

Zenzes, M.T., Krishnan, S., Krishnan, B., Zhang, H., and Casper, R.F. (1995) Cadmium accumulation in follicular fluid of women in in vitro fertilization-embryo transfer is higher in smokers, *Fertil. Steril.*, 64, 599–03.

chapter 8

Epidemiological and Occupational Studies of Metals in Male Reproductive Toxicity

Wendie A. Robbins
UCLA Center for Occupational and Environmental Health

Contents

Introduction

It is important to determine the effects of exogenous metal exposures on human male reproductive health. Recent reports of regional differences in human fertility, declining semen quality, and increasing incidence of abnormalities of the testis[1-9] have prompted research on environmental contaminants as potential etiologies, particularly estrogens and antiandrogens. Metals have received comparatively less attention for their potential effects on male reproductive health in spite of extensive use in modern society and wide distribution geographically. Workplaces in the United States also have pressing reasons to determine the effects of metals on male reproductive health. A 1991 Supreme Court decision in *International Union v. Johnson Controls*[10] involving workplace lead exposure set precedent for employer provided risk information related to reproductive hazards. Employers can be held liable for reproductive injuries incurred at work and are required to provide employees with scientifically based reproductive risk information. Liability extends to affected offspring (section 5a [1] 1970 OSH [Occupational Safety and Health] Act, the "general duty clause") suggesting research should include male-mediated effects of metals as well as direct effects on fertility or sexual performance.

A modest number of epidemiological studies have addressed effects of environmental and workplace metal exposures on human male reproductive health. Subsequent review papers repeatedly cite these findings even though the quality, generalizability, and conclusions of the published works vary (Table 8.1, Table 8.2). In spite of shortcomings, this body of epidemiologic literature has identified several metals and metal-working industries that demonstrate trends related to effects of environmental and/or workplace metals on male reproduction, for example, lead (reviewed in a previous chapter of this text), the occupational category "welder" (reviewed in this chapter), and alterations in gender ratios associated with boron (reviewed in this chapter). In addition, a fair amount has been published on constitutive medical conditions involving metals that effect male reproductive health, for example, hypogonadotropic hypogonadism related to idiopathic hemochromatosis or thalassaemia.[64-65] There is literature on metal micronutrients and their effect on male sexual health. For example, selenium has been shown to have positive effects on semen quality at some doses,[45,52,66] but impair human fertility at higher doses.[67] This chapter will not address the medical disease literature or micronutrient literature, and other than multifactorial genetic liabilities, concentrates only on effects of nonnutrient exogenous occupational or environmental metal exposures to otherwise healthy, free-living, human males.

Table 8.1 Selected Journal Reviews of Effects of Metals on Human Male Reproduction

Author	Metal Exposures	Outcomes
Antilla and Sallmen[11]	Pb, Hg, Ni, As, Cd, Cr, Zn Cu smelting	Spontaneous abortion
Baranski[12]	Pb, Mn, Hg, Antimonite Welding	Hormones, semen parameters, fertility, pregnancy outcomes
Figa-Talamanca et al.[13]	Al, Ca, Cd, Cu, Pb, Mg, Se, Zn	Semen parameters, fertility,
Lahdetie[14]	Pb, Hg, Cr, welder	Semen parameters, fertility, testicular atrophy
Lamb and Bennett[15]	Pb, Cd	Fertility, semen parameters
Narod et al.[16]	Pb-related employment	Spontaneous abortion, childhood cancer
Olsen et al.[17]	Metals	Childhood cancer
Olshan et al.[18]	Paternal occupation, i.e., welder, cutter, metal worker	Congenital anomalies
Robbins and Cousins[19]	Pb, Mn, Cd, Hg, B	Hormones, semen parameters, fertility, pregnancy outcomes, birth defects, childhood cancer
Rosenberg et al.[20]	Pb, Steel, Hg, Mn, Cu smelting	Impotence, hormones, semen parameters, fertility, pregnancy outcomes, sex ratio, birth defects, childhood cancer
Savitz and Chen[21]	Paternal occupation, i.e., welder, cutter	Childhood cancer
Savitz et al.[22]	Agent (Hg, Pb), industry (i.e., metallurgical, Cu smelter), occupation (i.e., dentistry)	Spontaneous abortion
Sheiner et al.[23]	Pb, Cd, Cr, Cu, welding	Semen parameters, testicular atrophy, pregnancy outcomes, birth defects
Steeno and Pangkahila[24]	Pb, B	Fertility, sexual performance, semen parameters
Tas et al.[25]	Cd, Mn, Pb, Hg, B	Hormones, semen parameters, fertility, pregnancy outcome, birth defects, childhood cancer

Methodological Considerations in Studies of Human Male Reproduction and Metals

Outcome Measures

The effect of metals on human male reproduction is generally assessed in epidemiologic studies through physical exam, sexual development, hormone

Table 8.2 Human Field Studies of Specific Metals and Indicators of Male Reproductive Health (Table entries indicate the numbers of studies with positive (+) or negative (−) findings concerning an association between the metal and the endpoint)

Reproductive Outcomes	Boron	Cadmium	Magnesium	Manganese	Mercury	Nickel	Selenium	Publication Reference Number for This Table
Reduced Fertility	1+ 2−	1+ 3−	4−	2+ 2−	3−	1+ 1−	1−	B 26,27,28/ Cd 31,33,34,43/ Mg 43,46,48,53/ Mn 30,31,51,61/ Hg 30,40,41/ Ni 55,61/ Se 46
Hormonal Imbalance		1+ 1−		1+ 1−	1+	1−		Cd 35,54/ Mn 42,62/ Hg 39/ Ni 62
Sperm Abnormalities								
General	1+	2+	5−	1−	1+	1+ 1−	1−	B 26/ Cd 32,34,44,45,54,57,60/ Mg 46,47,48,53,59/ Mn 62/ Hg 38,50,57/ Ni 58,62/ Se 45,46,49,52
Count						1+	2−	
Motility		4−					2−	
Morphology		3+			1+			
CASA					1+			
Abnormal DNA/ Chromatin								
8-OHdG		1+					1−	Cd 45/ Se 45
Poor Offspring Health								
Pregnancy Loss		1−			2+	1+		B 27,29,63/ Cd 37/ Hg 36,40,41/ Ni 55,56
Gender Ratio	3+				1−	1+		

levels, specific characteristics of ejaculated semen, fertility and other aspects of reproductive history, and health of offspring (Table 8.3). Most studies incorporate more than one outcome because the process of reproduction is complex, and information on a single measure may fail to detect toxic effects.[68] Choosing the most informative outcomes for study is best guided by knowledge of the toxicologic mechanisms of the metal of interest. For example, lipid peroxidation of sperm membranes might be an informative outcome in studies of metals that change oxidation states as a function of their toxicity, such as nickel, manganese, or chromium, but not as informative in studies of aluminum, cadmium, or lead.[69] For lead toxicity, chromatin condensation in ejaculated sperm cells could be informative as an outcome because lead is hypothesized to displace zinc in protamine, thus interfering with chromatin remodeling during spermiogenesis. Studies of toxicologic mechanisms in rodents, cell lines, and other model organisms, as described in earlier chapters in this text, inform choices in the outcome measurements used in epidemiologic designs.

However, there are two major caveats. First, knowledge of differences between species is critical because reproductive physiology is diverse among mammals and can limit extrapolation between species. For example, variation in sequence and ratio of protamine 1 and 2 across species may modify the effect of certain metals on sperm condensation and limit utility of comparisons of sperm chromatin outcome measures between humans and animals. Second, although biologic plausibility is advantageous, delineation of specific toxicologic mechanisms is not required for epidemiologic inquiry.[70] An illustration of this related to metal toxicity is the comprehensive body of work on welding and male reproductive effects. Much of the early work with significant findings was contributed by Jens Peter Bonde and associated researchers in Denmark[71] and preceded complementary work in animals. There are multiple examples in the general public health literature where significant findings in epidemiologic studies drove science and public policy prior to confirmation of specific underlying toxicologic mechanisms (e.g., cigarette smoking and lung cancer, water fluoridation, and dental carries).

Biologic Outcome Markers
Seminal fluid and sperm cells in ejaculated semen have been used to assess the effects of metals on male reproductive health in a number of epidemiologic studies (Table 8.2). To the credit of researchers and peer reviewers in this field, published metal studies using ejaculated semen as an outcome measure have generally conformed to accepted criteria for collection, processing, analysis, and reporting of seminal fluid characteristics and sperm parameters as outlined in the *World Health Organization (WHO) Laboratory Manual for the Examination of Human Semen and Sperm-Cervical Mucus Interaction*, editions 1 to 4.[72] The four editions of the WHO manual span the years 1980 (1st edition) through 1999 (4th edition). Many investigators with publications describing male reproductive effects from exposure to metals have

Table 8.3 Male-Specific Endpoints of Sexual Function and
Fertility That Have Been Used in Published Metal Studies

Biological Markers	Nonbiological Markers
Seminal Fluid	*Reproductive History*
Volume	Infertility
Viscosity	Time to pregnancy
Parent metal compound	Spontaneous abortion
Metal intermediates	Stillbirth
LDH	Numbers of offspring
LDH-C$_4$	Gender ratios at birth
Zinc (normative measure)	Sexual performance
	Erectile dysfunction
	Libido
Sperm Cells	*Offspring Health*
Count	Birth weight
Morphology	Birth defects
Head	Childhood cancer
Midpiece	Developmental delays
Tail	
Vitality	
Motility	
CASA	
VCL (curvilinear velocity)	
VSL (straight line velocity)	
LIN (linearity of motion path)	
VAP (average path velocity)	
ALH (amplitude of lateral head displacement)	
DNA and Chromatin integrity	
8-OhdG	
Chromatin decondensation	
Blood	
FSH, LH, Testosterone	
Parent metal compound	
Metal intermediates	
Physical Exam	
Normality of genitalia	
Androgenized male	
Testicular size	

taken great care to standardize semen analysis procedures to the best andrology practice of the corresponding time era.

The WHO manual lists reference values for semen variables; these have
changed slightly over the years but currently, for the parameters used most
frequently in semen field studies, they are: volume 2.0 ml or more, pH 7.2–8.0,

sperm concentration 20×10^6 spermatozoa/ml or more, total sperm count 40×10^6 spermatozoa/ejaculate or more, vitality 50% or more, motility 50% or more, morphology 30% or more with normal forms adjusted downward for strict criteria, and zinc 2.4 µmol/ejaculate or more. Prevalence of abnormal semen parameters in the base population at risk affects the power and choice of study design. For case control studies, relatively common events need a larger sample size with a large number of cases (e.g., history of erectile dysfunction reported to occur at a frequency of greater than 30%); rare events need smaller samples of cases, but since they are rare, need a larger population at risk in order to obtain the smaller number of cases (e.g., azoospermia reported to occur at a frequency of 1%).

Semen parameters can be used to indicate direct toxic effects on the male reproductive system (e.g., reduced sperm count or motility subsequent to metal exposure) or to predict adverse effects on fertility or offspring health. Currently it is estimated that 15% of U.S. couples suffer infertility with 35 to 40% due to male factors. Even though reduced sperm numbers, decreased motile vigor, and abnormal morphology are generally associated with decreased fertility, there is no definitive threshold below which a pregnancy is impossible. For example, Jounannet et al.[73] and Bonde et al.[74] have demonstrated that sperm concentrations below 20×10^6 spermatozoa/ml correlate with infertility, although Horvath et al.[75] and Chia et al.[76] have shown that pregnancies do occur in populations with sperm counts well below this level. Because of this, it is necessary to consider that metal exposures can appear to affect semen parameters to the point of infertility, but an individual sperm carrying metal-induced defects may still fertilize and produce offspring.

Computer-aided semen analysis (CASA) has been informative in a number of epidemiologic studies looking at metals. CASA outcomes that seem the least dependent upon instrument brand or software version, and thus are most comparable across epidemiologic field studies, are straightline velocity (VSL), an indication of sperm progression; curvilinear velocity (VCL), an indication of sperm vigor; and linearity (VSL/VCL × 100) or LIN, an indicator of the straightness of the sperm track. The increasing portability of computer-aided semen analysis equipment has led to increasing incorporation of computer-aided assessments in semen field studies. A benefit is greater standardization of laboratory assessments in the field and the ability to store images for further critical analysis at a later date by researchers in a central laboratory.

To date, the semen quality outcomes utilized most often in investigations of metals on male reproduction have been count, motility, and morphology. Used together these parameters provide a good general assessment of male fertility and reproductive health status. All are relatively easy to assay in an ejaculate and can be accomplished in field studies. However, the science of andrology is constantly developing new techniques to measure aspects of sperm that predict male fertility and reproductive health. Some examples include contemporary DNA integrity assays, such as the single cell gel electrophoresis assay (Comet assay), which detects DNA strand breakage in single sperm cells; fluorescence *in situ* hybridization (sperm-FISH), which detects abnormalities in chromosome number or structure in individual sperm cells; sperm chromatin structure assay (SCSA), which detects DNA alkaline-sensitive sites in sperm populations; and the terminal deoxynucleotidyl transferase (TdT) nick end-labeling assay (TUNEL), which can be used to detect sperm DNA fragmentation in either slide-based or flow cytometry applications. These assays are being increasingly applied in sperm toxicology studies because they improve sensitivity to detect targeted toxic effects.[77] Recent examples of sperm DNA chromatin assays that have been utilized to look specifically at associations between metals and sperm health include correlations between seminal fluid zinc and SCSA,[78] and correlations between seminal plasma cadmium, lead, and selenium with the oxidative DNA-adduct 8-hydroxydeoxyguanosine (8-OhdG) in sperm as a marker of oxidative stress.[45] An important attribute of some DNA chromatin assays is that the types of damage being measured can be transmitted to offspring (e.g., sperm aneuploidy or DNA alkaline-sensitive sites), which allows estimation of toxic effects independent of the female partner's reproductive health.

Hormones have been included in a number of occupational epidemiological investigations of metals, although most of the metal reproductive toxicants identified to date appear to act primarily through direct damage to sperm cells or seminiferous tubules and only secondarily through hormone profiles. In andrology clinic populations, hormone evaluations would be part of the routine workup, and this data would be accessible to researchers. However, at the population level, human male gonadotropin levels have not been as informative as other measures in investigating the toxicity of metals on reproduction (Table 8.2).

Nonbiological Markers

Fertility and Pregnancy History. Among married couples in the United States, 7% report inability to conceive after 12 months of unprotected intercourse, and 15% report a past infertility-associated health care visit.[79] Inability to achieve a pregnancy after 1 year of unprotected intercourse is the usual definition for infertility used in epidemiologic studies, and it has been assessed as a dichotomous outcome in a number of occupational epidemiologic studies of metals. The researcher simply asks the question of men or their partners using questionnaire or interview formats. If answered "yes," details may be obtained regarding whether the couple has seen a clinical provider, whether the etiology of infertility has been determined (male or female partner), or the history of previous semen analysis and results. In metal studies, populations have also been identified through infertility clinics. This is a potential problem in terms of generalizability of findings, but clinic populations are already providing semen, blood specimens, historical reproductive data, and medical and physical exam data. In some cases, males attending infertility clinics will be grouped as to male factor or female factor infertility and compared on occupational or environmental metal exposure, or alternatively compared on the metal exposure of interest to healthy men of proven fertility donating semen to insemination programs.

Fecundity is a couple's probability of pregnancy in a noncontracepting menstrual cycle. It is measured in epidemiology studies as the number of menstrual cycles it takes a couple with regular, unprotected intercourse to achieve pregnancy or *time to pregnancy* (TTP). Rather than indicating impairment in a specific biologic pathway, TTP, like a history of infertility, reflects the ability of the entire reproductive system and has been associated with increased risk for low birth weight and preterm delivery of offspring.[80–82] TTP has been used in a number of epidemiologic studies of lead exposure to males, welding exposures,[61] metal fume exposures in minting,[55] and zinc, magnesium, and calcium exposure in couples belonging to trade unions.[48] TTP has been shown to be a sensitive measure of fecundity in couples even in studies requiring long recall periods and retrospective assessment through questionnaires. It has been validated for occupational groups and in data collected solely from the male partner.[83] Stable estimates of population-based TTP distributions can be achieved with 200 to 300 pregnancies. Tingen et al.[84] and Lamb and Bennett[15] present discussions of methodological issues, statistical modeling, and potential biases in prospective and retrospective designs using TTP that are helpful when planning studies of fertility related to exogenous exposures.

Numbers of progeny compared between groups is another outcome measure that has been used in studies of metals and male reproductive health. For example, Gennart et al.[31] compared the probability of live-born offspring of cadmium-, lead-, and manganese-exposed workers before and after the onset of exposure versus unexposed controls. Whorton et al.[85] used standardized fertility ratios for the United States to demonstrate boron workers were more fertile than U.S. males in the general population. Standardized rates for a nation lack individual-level data on confounding or other factors (e.g., smoking, socioeconomic status, access to health care, area-specific family size norms) and can lead to bias, although this type of comparison is often used as a first pass to suggest areas for future research.

Ratio of male to female live births has been used in epidemiological studies investigating effects of metals on male reproduction. Figa-Talamanca and Petrelli[56] reported a statistically significant reduction in the expected ratio of male to female births to workers exposed to metal fumes (nickel and chromium) in an Italian mint. The workers categorized as most highly exposed based on job title (n = 63 founders) reported 33.3% male births for all their children compared to 58.3% for unexposed workers (n = 48 administrators) and 30.2% for the most recently born child compared to 57.1% for the administrators. Three epidemiologic studies have documented the phenomenon of gender ratio deviations related to boron exposures although the differences did not reach statistical significance. Whorton et al.[85] reported an excess of female births to 542 boron workers at a U.S. boron mining and processing facility (52.7% female births compared to 48.8% in the U.S. population). Sayli[63] reported a reduction in the expected number of male births to residents living and working on or near the boron ore beds in Turkey (49.6% males compared to 51.9% males for the population of rural Turkey not exposed to boron). Her findings were based on 2438 births including live and stillborn. Robbins et al.[29] found a reduction in the expected ratio of male to female births to 828 married boron workers in northeastern China (1.07 for boron workers compared to 1.18 for controls and 1.15 for China as a nation). In all three of these studies, the male-to-female birth ratios of offspring were based on interview or questionnaire data. Gender ratio is easily accessible and thus a cost-effective indicator useful in epidemiologic studies of toxic effects of metals if there is sufficient sample size. However, just as with using standardized fertility data for a nation, population-level data on male-to-female birth ratios lacks individual level data on confounding.

Spontaneous abortion has been evaluated in a number of studies looking at metals and male reproductive health. Pregnancy loss can be difficult to measure because couples can lose a pregnancy before they even realize a conception has occurred. Spontaneous abortion has been measured in relationship to paternal exposures to lead in multiple investigations, but also to boron, cadmium, mercury, and nickel. Among clinically recognized pregnancies, approximately 20% will be spontaneously aborted, while it is estimated that approximately 50 to 80% of total human conceptions will be lost prior birth. Failure to recognize or consider in the analysis very early fetal or even embryonic deaths may lead to failure to detect important exposure effects.[86] In retrospective designs, the measure of pregnancy loss is usually underreported,[87] and less reliable in terms of timing in relationship to exposure. Of importance, chromosomal abnormalities are estimated to occur in 30 to 40% of spontaneous abortions[88] so this outcome could be informative in investigations of metals that interfere with aspects of cell cycle, mitosis, or meiosis.

Offspring Health.   Of live births in the United States, approximately 7% are of low birth weight and 2 to 3% will have other birth defects obvious at birth.[88–90] Spontaneous genetic mutations and chromosomal aberrations play a large part in birth defects, but the cause of greater than 40% of malformations at birth remains unknown.[91] Overall, chromosomal abnormalities are estimated to occur in 0.2% of live births.[88] Paternal contributions to chromosomal abnormalities in those who are live born is significant, particularly regarding the sex chromosomes (~50% for XXY, 100% for XYY, and up to 80% for XO). This suggests that investigations of occupational or environmental metal exposures of men, which might affect sperm chromosome number and be transmitted to offspring, are important areas for study.

Exposure Characteristics

Exposure assessment is critical to determining adverse effects of metals on human male reproduction. Historically, this has been a weakness in epidemiologic studies of metals, often leading to conflicting findings across investigations. Combining traditional exposure measurements (e.g., job title or industrial hygiene records) with measurement of metal species in biological samples can improve exposure assessment and help clarify associations. For example, Danadevi et al.,[58] using job history to categorize welders and controls, found significantly poorer sperm count, motility, morphology, and vitality in welders. However, when semen parameters were regressed specifically on levels of nickel and chromium in blood, sperm tail defects, progressive motility, and vitality were correlated with nickel, whereas decreased count was correlated with blood chromium (Table 8.4). Thus, the study confirmed a trend in the literature reporting adverse effects of welding on male reproduction, but demonstrated a dose-related effect of nickel and chromium. It is interesting to note that nickel and chromium explained only a portion of the variation in semen effects detected, which suggests other factors must also be involved, for example, heat.[92]

Timing of Metal Exposures

The male reproductive system has sensitive windows for toxic exposures both during development and throughout adult life. Metals can cause different effects depending on the timing of exposure. Exposures *in utero* or during childhood and development may lead to later reproductive failure. Exposures to the adult may have direct effects on gamete production, spermiation, maturation of sperm in the epididymis, ejaculation, accessory organs, seminal fluid, and the testis, and subsequently hormones and sexual performance. In addition, exposures to the adult male may affect offspring through exposing the mother and fetus to contaminated semen or transmission of abnormal sperm DNA/chromatin to progeny. For example, Buckley et al.[93] used case-control design and retrospective questionnaire exposure data to show that paternal occupational exposure to metal dusts during prenatal development of offspring was associated with childhood hepatoblastoma (OR 3.0, 95% CI 0.9 to 13.5) and Wilkins and Koutras[94] used case-control design and birth certificate job title data to show increased risks for brain cancer in the offspring of men working in the general category of metal industries at the time of the birth of the child (OR 1.8, 95% CI 1.1 to 2.9). In the Wilkins and Koutras analysis,[94] the occupational subgroups including welders (OR 2.7), metal processing/ore refining (OR 5.0), and foundry (OR 5.9) all suggested increased risk to offspring, although subgroup numbers were too small to reach statistical significance.

Spermatogenesis involves one of the most rapidly proliferating tissues in the body. Timing from spermatogonia precursor mitotic division yielding two type A spermatogonia (dark and pale), through meiosis, which lasts more than three weeks (spermatocytes), culminating in haploid cells (spermatids), has been estimated for the human at approximately 70 days.[95-96] Damage from metal exposures early in the spermatogenic cycle during spermatogenic cell differentiation might cause long-term disturbances in semen quality, whereas damage during spermiation or to sperm resident in the epididymis would result in more transient effects. It is important to recognize that cell transit time through the epididymis is variable in humans and highly

Table 8.4 Selected Epidemiologic Studies on Male Reproductive Effects from Exposure to Welding. OR = odds ratio, CI = 95% confidence interval, NS = not significant

Design	Exposure Assessment	Timing of Exposure	Reproductive Outcomes	Control for Confounding	Reference
Case-Control Infertile couples n = 1069 Fertile couples n = 4305	Questionnaire on work completed by female partner	Exposures preceding care for infertility or childbirth	Semen quality, conception within one year Findings: Poor semen quality stainless steel welding OR 1.7 CI 0.9–2.9 Delay conception Nonstainless steel welding OR 1.4 CI 1.1–1.8	Maternal and paternal factors	Rachootin and Olsen[156]
Case-Control Infertility Clinic Case defined by semen analysis: Conc <20 × 10^6 /mL Motil <50% Morph<50% (Samples collected 1981–1983)	Questionnaire on work complete by the worker Exposure Groups: Welders Metal workers Other industrial No chemical or physical exposures	6 months prior to conception	Semen quality Findings: Poor motility Group 1 vs. Group 4 OR 2.0 CI 1.2–3.5 Group 2 vs. Group 4 NS	Fertility of wife, age	Mortensen[153]
Cross-sectional stainless steel welders n = 77 Nonwelders n = 68 from the same plants	Job tasks Blood cadmium	Cross-sectional	Semen quality	Age, smoking	Jelnes and Knudsen[155]

Study population	Exposure assessment	Timing	Results	Confounders	Reference
Cohort of stainless steel or mild steel workers n = 3702 (Cohort assembled 1964–1984)	Questionnaire on work history used to calculate person years of exposure and type of welding. Exposure Groups: Mild steel welders and stainless steel welders. 0 = unexposed 1 = stainless steel, tungsten inert gas 2 = stainless steel metal arc (Wendy are the mild steel workers Group 3?)	1 year prior to birth of child	Births recorded in the Danish birth register. In years exposed to welding vs. unexposed years OR 0.89 CI 0.83–0.97. Stainless steel OR 0.98 CI 0.70–1.37 irregardless Group 1 or 2. Mild steel OR 0.86 CI 0.76–0.99	Alcohol, smoking, birth cohort, age parity, previous year occupation	Bonde et al.[154]
Prospective cohort. Six Denmark workplaces Stainless steel welders n = 30 mild steel welders n = 30 non-welding metal workers plus electricians n = 47	Interview on occupational exposure, personal habits, urogenital disorders. Urinary and blood chromium	Semen samples once per month × 3 months. Two urine and blood samples at the end of the 3 months	Semen analysis and blood FSH, LH, testosterone NS differences across the groups on outcomes NS association between biological fluid chromium and outcomes	Age, smoking, alcohol, spillage, hot baths, fever, shift work	Bonde and Ernst[141]

Table 8.4 Selected Epidemiologic Studies on Male Reproductive Effects from Exposure to Welding. OR = odds ratio, CI = 95% confidence interval, NS = not significant (continued)

Design	Exposure Assessment	Timing of Exposure	Reproductive Outcomes	Control for Confounding	Reference
Prospective cohort Denmark couples at time discontinue birth control Welders n = 126 Nonwelding metal workers n = 68 Nonmetal workers n = 200 (Cohort enrolled 1992–1994)	Questionnaire on details of welding type, engineering controls, duration, other toxic exposures Urinary chromium, nickel, manganese pre- and postshift	Periconception	Semen analysis with CASA Blood FSH, LH, testosterone/SHBG NS differences between the groups NS differences in urinary chromium, nickel, manganese between the groups and between pre- and postshift urines	Age, abstinence, season, center, smoking, alcohol, spillage	Hjollund et al.[61] Hjollund et al.[62]
Case-control Infertility clinic Case = metal worker / welder n = 20 Control = other infertility patients n = 1382	Questionnaire on work and state statistics on workers in the population base.		Semen analysis Findings: Poor motility OR 5.99 CI 1.38–26.00	Age, abstinence	Kenkel et al.[157]

Cross-sectional 57 welders exposed for 2–21 years 57 controls not exposed to welding Welding plant in south India	Questionnaire on work exposure n = 28 welders, 27 controls blood nickel and chromium	Cross-sectional	Medical exam reproductive history, semen parameters Findings: Decreased semen quality in welders compared to controls $p < 0.001$ Nickel correlated with motility, tail defects, vitality $p < 0.04$ Chromium correlated with count, motility, tail defects, vitality $p < 0.02$	Age, smoking, medicine usage, exposure to other repro toxins	Danadevi et al.[58]

susceptible to individual characteristics, such as nutrition, hydration, illness, and frequency of ejaculation. Because this can complicate appropriate timing of semen collection in field studies, researchers generally plan collection that encompasses exposures of interest during a full cycle of spermatogenesis (90 days). In the case of a known mechanism associated with a specific stage of sperm development, the sampling can be targeted to ensure maximal probability of capturing that effect. In well-conducted semen field studies, researchers control for ejaculation frequency related to the research specimen of interest. If the intent is to look at potential contamination of the maternal-fetal unit through contaminated semen, timing of specimen collection is based on toxicodynamic information for both male and female partners. Movement into, and duration of, the metal or metal intermediate in seminal plasma following exposure, as well as time the element would remain in the female genital tract, would be taken into consideration.[97]

Trying to reconstruct exposures for specific time windows of the spermatogenic cycle is difficult in the human. Variable transit time through the epididymis is especially problematic. In the case of chronic environmental or workplace exposures it is likely they will be highly correlated across all time windows of sensitivity. Cumulative effects are usually ascertained in workplace studies by correlating adverse reproductive outcomes with duration of exposed work, taking into account time elapsed since last exposure. Designing acute exposure studies can be more straightforward, but is still much less precise than can be achieved with animal work.

Characteristics of the Exposure

In animal studies or other model test systems, exposures are carefully constructed and delivered systematically. This is not the case in human field studies. Even something as obvious as route of exposure, which seems obvious in animal studies, becomes complicated in human investigations. For example, in an occupational study of inhalable workplace dust containing metal, other sources of the same metal in food, water, and medicinals need to be considered or conclusions based on occupation alone would not accurately reflect exposure. Most of the current literature on metal exposure and male reproductive effects is weakest in terms of exposure assessment. This is particularly distressing because the same metal can have varying effects depending on speciation (e.g., inorganic and organic), dose, timing, and duration of exposures.

Male Characteristics that Affect Exposure and Outcome Relationships

A number of characteristics have been identified that influence male reproductive health and could impact studies investigating the effects of metals. These factors are potential confounders and need to be assessed in study subjects. They are considered in initial study design and sampling, and are used to power the investigation to be able to ascertain associations and interactions. These factors will also need to be incorporated in interpretation and dissemination of findings.

Lifestyle factors such as use of alcohol, caffeine, and cigarette smoking have been evaluated in numerous male reproductive studies with enough cumulative evidence to merit their consideration as potential confounders for most male reproductive outcomes of interest. Smoking has been associated with decreased fertility, poor semen quality, hormonal changes, and effects on sperm DNA and chromatin.[98–104] Alcohol appears to have little effect on fecundability or semen quality, except at high or daily doses,[101,105–107] but has variable effects on sperm aneuploidy across doses.[102,108] Thus, in studies of direct cytogenetic endpoints or outcomes associated with cytogenetic abnormalities, such as pregnancy loss or birth defects, it is important to include measures of alcohol exposure. Caffeine intake of more than 700 mg per day has been associated with decreased fecundability,[106] even though lesser amounts have been reported to have little effect,[101,109] and there is a single report of dose-dependent effects on sperm aneuploidy.[102]

Tight trousers, hot baths, hot work areas, or activities related to increasing testicular heat, such as extended periods of sitting or car driving, could affect semen quality and male reproductive outcomes.[110–111] Seasonal changes in semen parameters and sperm chromatin condensation have been reported by enough researchers to warrant consideration in study design.[112–116]

Medications such as antineoplastic agents, estrogens, gastrointestinal medications, or exposure to x-rays have been correlated with decreased sperm counts, changes in sperm motility, or aneuploidy in sperm.[117–119] Medical conditions such as past or present varicocele, recent infection, high fever, testicular injury, kidney infection, or diabetes have been shown to lower sperm concentration, total sperm count, motility, and morphology, as well as affect sperm chromatin structure.[119,120] Medical history information is usually ascertained through questionnaire, but in some settings can be taken from clinical records.

Increasing age has been associated with decline in semen quality.[116] This is especially relevant to workplace studies in industrialized nations where the workforce is aging. Because aging itself leads to decreased metabolic ability, decline in kidney function, and accumulated mutations, males may be more sensitive to metal exposures when they are older. Most of the currently published epidemiological literature describing metal exposures related to male reproductive health concentrate on adult, nongeriatric males.

Dietary practice is particularly important in studies of metals and reproduction because a number of metals are essential micronutrients, e.g., chromium, cobalt, manganese, selenium, and zinc.[121] In addition there is interdependence of a number of metals in molecular mechanisms related to sperm function, e.g., magnesium and zinc.[48,53] Zinc is particularly important in sperm health and is essential for sperm membrane fluidity, stability of chromatin, acrosome reaction, and is correlated with seminal fluid citric acid and fructose.[122,123] It has been suggested that nonessential metals like cadmium (a contaminant in some foods) may compete for binding sites on membrane proteins involved in the transport of zinc and other essential elements (such as calcium and iron) resulting in toxicity.[124]

Geographic location has been associated with variation in sperm quality and male fertility. Whether this relationship is due to environmental exposures, ethnic differences, or genetics is not clear. However, regional differences explain much of the variation in semen quality and fertility worldwide and, therefore, is an important consideration in study design and comparison of study findings across regions.

Genetics and epigenetic factors are also important considerations in investigations of metal toxicity on male reproduction. Heterozygotes for autosomal recessive metal metabolism and storage diseases (e.g., Wilson's Disease) and genetic variability in detoxification of metals (e.g., metallothionein expression)[125] may result in reduced enzyme activities and hypersensitivity to metal exposures. Multifactorial genetic liabilities are common in man[126] and a fraction of subjects in any population-based epidemiologic study are likely to be hypersensitive. This could directly affect reproductive outcomes being studied or impact health effects generally and indirectly affect reproductive health.

Psychological factors such as stress, mood, and self-esteem have been associated with various aspects of male reproductive health, including semen quality (as reviewed by Sheiner et al.[23]). As modern workplaces reduce chemical and physical toxicants by successful engineering and other controls, psychological aspects of work that affect health are gaining attention. Interactions between metal exposures and job strain are likely to be incorporated in more studies as the tools to measure workplace stress are refined.

Historically, epidemiologic studies of metals and male reproductive health have adjusted for some of the potential confounders listed above, as well as other known reproductive toxicants involved with hobbies. Other characteristics that affect male reproductive health would be expected to remain relatively stable across changes in exposure status (e.g., attitude toward family size, sexual behavior, contraceptive practice, and urogenital disease), and are not commonly measured. At a minimum, age and smoking have been addressed in many of the studies of metals and male reproduction listed in Table 8.2.

In summary, timing of exposure and potential confounders, as well as measurement of more than a single male reproductive outcome need to be evaluated in order to accurately capture specific effects of metals on male reproductive health. Exposure assessment has been the weakest point of the majority of epidemiologic studies in the past, but careful attention to timing of exposure and use of markers of internal dose are helping to improve this. A number of potential confounding or interacting variables between metals and their effects on male reproduction have been identified with enough consensus to merit their measurement and inclusion in any future investigations.

Study Design Issues for Studies of Metals and Male Reproduction

Study design incorporates characteristics of outcome measures such as deciding on dichotomous versus continuous measures and deciding on metal species and biologic matrices to be evaluated, in addition to considerations of sensitivity and practicality of choices to measure particular outcomes. Design also incorporates characteristics of the exposures, for example, route, duration, and timing in relationship to the reproductive outcomes being evaluated. Study design also considers potential confounders and bias. A number of potential confounders that might be responsible for differences in reproductive health seen in men exposed to metals of interest were discussed in the previous section, "Male Characteristics that Affect Exposure Outcome Relationships." Generally studies will be designed and powered during planning to allow control of these extraneous factors in the analysis if they are determined to be true confounders. Attention to potential bias is always a concern in observational studies. A single study reporting a weak association between a specific metal exposure and male reproductive effects is more likely to be explained by undetected biases than a study finding a strong association or replicating previous observations using different population groups.[127]

Semen parameters are frequently used as continuous outcome measures and may be correlated with continuous measures of metal exposure. Because semen parameters are not normally distributed, transformations are used in order to apply parametric statistics for hypotheses testing. Results can be hard to interpret, especially if statistically significant findings are detected for differences of questionable biological significance. To deal with this, some researchers use case-control designs in male reproductive health studies. Problems with case-control designs arise first when defining a case. Will case definitions be based on one or more outcome parameters, what parameters will be used, and what values will be used to define the parameters? Clinically defined parameters may not perform well with data sets being used to detect associations with low or moderate metal exposures. For example, Bigelow et al.[119] were able to demonstrate loss of power to detect significant correlations between semen parameters and the occupation of welder when men were dichotomized as normal versus nonnormal on the basis of semen parameter cut points. In case-control studies, choice of comparison group is also critical. Although donor bias has often been ascribed to semen studies, it is probably overstated. Lamb and Bennett[15] showed that sperm counts were similar in healthy men requesting vasectomy, nonmale factor infertility patients, factory workers, and agricultural workers.

When reproductive outcomes of interest are rare (e.g., birth defects or childhood cancer) case ascertainment often depends on registries or birth certificate data. Registries exist for birth defects and childhood cancer in most developed nations, but quality and coverage varies. Taskinen[68] recommends obtaining and reporting coverage, sensitivity, and accuracy of the registry or database when reporting findings. An excellent example of use of a cancer registry to look at paternal exposures related to childhood cancer is the study by De Roos et al.[128] reporting an increased risk for neuroblastoma in offspring of men exposed to solders (OR 2.6, 95% CI 0.9 to 7.1). Birth certificate data and registry data varies in quality and utility for research studies. Registries demonstrate a spectrum of detail. For example, the registry for Assisted Reproductive Technologies (ART) includes detailed data on reproductive procedures in relation to reproductive cycles, pregnancy losses, and all major and minor birth defect outcomes, whereas other formal registries in the United States include only major birth defects. As suggested by Marcus et al.[129] there is a need for establishing a common, uniform data set in the United States that captures reproductive outcomes such as spontaneous abortion and infertility. This would facilitate the understanding of potential hazards to reproductive health from metals and other environmental exposures.

Specific Metals and Human Male Reproduction

Human field studies of specific metals and biologic indicators of male reproduction are summarized in Table 8.2. This table gives the numbers of studies with positive and negative findings. Selected studies are discussed in more detail below.

Boron

Borates are widely used in industry for the production of glass, fiberglass, detergents and bleaches, enamels, glazes, metals and alloys, fire retardants, jet fuels, pesticides, paper, soaps, and cosmetics.[130,131] Some of the highest exposures occur during mining and refining ore into boric acid, borax, and other borates. Boron may be an essential human micronutrient, although high concentrations are toxic. A team from government, industry, and labor identified boric acid as one of four chemicals with the highest priority for occupational health field study among 43 National Toxicology Program (NTP) animal reproductive toxicants.[132] Selection was based on data from Reproductive Assessment through Continuous Breeding (NTP/RACB) protocols showing a Lowest Observable Effect Level (LOEL) less than 250 mg/kg/day for reproductive effects, estimates of greater than 100,000 workers exposed based on the National Occupational Exposure Survey (NOES), and limited epidemiological data.[132] Data on human reproductive effects of boron have been published using four different population groups to date. In a 1972 Russian publication, Tarasenko et al.[26] reported testicular dysfunction and sterility in workers exposed to high levels of boric acid dust, but the exposure levels were not reported. Whorton et al.[85] used historical exposure data and questionnaires to obtain information on live births to 542 male workers in boron mining and borax production at U.S. Borax in the Mojave Desert, California. Standardized Birth Ratios (SBRs) were used to demonstrate no evidence of adverse effects of borates on fertility, although a perturbed male–female ratio in offspring was noted (more females than males). Sayli[28,63] studied families in Turkey living in regions with high risk for boron exposure (communities around boron mines or processing plants, or workers in the mines) and to persons living in regions expected to have low boron exposures. Questions on numbers of pregnancies, early infant deaths, congenital malformations, stillbirths, and spontaneous abortions were asked. No adverse effects on either male or female fertility or development were found, although more female than male offspring were born in the highly exposed regions. In preliminary findings by Robbins et al.,[29] based on research in northeastern China using community-based participatory research methods and interviews with 1187 boron workers and controls, less than the expected ratio of male to female children were born to boron workers compared to controls and compared to male-to-female birth ratios for China as a nation. Thus, in all three human study groups where gender ratios at birth were evaluated, regardless of study design, disturbance in expected male-to-female ratios were found, although findings did not reach statistical significance.

Cadmium

The two major routes of cadmium exposure for humans in the general population are inhalation, primarily as CdO, and ingestion, primarily as $CdCl_2$ in contaminated water or food.[124] Cigarette smoke is the largest source of cadmium exposure outside of occupational settings. A number of studies have investigated blood or seminal fluid levels of cadmium in relationship to semen quality, as well as looking at baseline cadmium levels in ejaculated cells. Battersby et al.[133] evaluated ejaculated sperm cells and could not detect cadmium in healthy sperm cells. After incubation of the cells with cadmium chloride, an effect on motility was found: approximately 30% reduction in motility during two hours of incubation with 0.02 mM cadmium chloride and approximately 35% reduction at 2.0 mM cadmium chloride compared to control. Chia et al.[32] showed statistically significant trends between cigarette-years, blood cadmium, seminal fluid cadmium, sperm density, and morphology, supporting the need to include current smoking as well as smoking history in field studies of cadmium and male reproductive health. Telisman et al.[54] reported a statistically significant correlation between blood cadmium and percentage of morphologically abnormal sperm ($r = 0.158$, $p < 0.05$), serum LH ($r = 0.185$, $p < 0.05$), testosterone ($r = 0.295$, $p < 0.005$), and serum prolactin ($r = -0.168$, $p < 0.05$) in 98 lead workers plus 51 controls with a range of blood cadmium from 0.16 to 13.33 µg/l. The authors found that the greatest contributor to cadmium in blood or seminal fluid was smoking. Hovatta et al.,[60] enrolling 27 rural men working in a refinery or polyolefin factory (mean blood cadmium 0.001 ± 0.001 mg/kg) and 45 urban semen donors (mean blood cadmium 0.003 ± 0.004 mg/kg), found no correlation with sperm count, motility, or morphology at these low cadmium levels, even restricting the linear regression to men having the highest blood cadmium concentrations (0.02 mg/kg). In a study of workers exposed to cadmium in two smelters in Belgian in 1988 and 1989, Gennart et al.[31] did not find a difference between live births to 83 exposed (urinary cadmium level geometric mean 6.94 mg/g creatinine) vs. 138 unexposed workers. Based on andrology clinic populations, effects of cadmium dose on semen parameters are mixed. Xu et al.[45] in 56 healthy nonsmokers with mean seminal cadmium levels of 0.78 µg/l found a statistically significant inverse relationship between cadmium and sperm density (r = -0.28, p < 0.05) but not motility or morphology. Chia et al.[57] evaluated 35 nonsmoking males for infertility in a Singapore clinic and found statistically significant correlations of blood cadmium (mean 1.35 ± 1.0 µg/l) with immature forms of sperm in the ejaculate (r = 0.45, p < 0.01) and with sperm midpiece defects (r = 0.42, p < 0.05), but no correlation with sperm density. Two additional studies failed to detect an association between cadmium in seminal plasma and fertility status,[34,134] whereas Umeyama et al.[43] reported infertile men had higher levels of cadmium in seminal fluid (mean 13 ± 12 µg/l) compared to fertile men (mean 5 ± 6 µg/l). In addition, there was one case report that did not find effects on fertility.[33] Although it is difficult to compare across the works due to reports of blood levels in some, urinary levels or seminal fluid levels in others, and inconsistent reporting on smoking, there are no significant findings in three studies looking at fertility outcomes except one at a very high seminal cadmium level, and no findings of effects on sperm motility other than *in vitro*. Two out of four studies reported significant findings related to sperm density and three out of four reported significant findings related to sperm morphology. Thus, it appears there is a dose-dependent effect of cadmium on seminal fluid parameters, particularly morphology. However, much higher human exposures are required before effects are seen on fertility.

Chromium

Human exposure to chromium and chromates occurs through air, food, water, cigarette smoke, and occupations such as welding.[71,135] Chromium levels have been repeatedly shown to be higher in human seminal fluid than in corresponding blood.[135,136] In nonhuman test systems, Cr(VI) compounds induce chromosomal aberrations, mutations in germ cells, DNA strand breaks, DNA protein cross-links, and protein modifications. Because chromosome aberrations have been associated with cancer risk,[137] it is postulated that Cr(VI)-induced chromosomal changes in sperm may be responsible for the increased cancer risk seen in offspring of fathers employed in the metal industry with chromium exposures, although associations are weak.[17,21,138-140] However, investigations of direct effects of Cr(VI) on human sperm DNA (e.g., Comet, TUNEL) have not yet been published.

Several epidemiological studies have looked for direct effects of chromium on semen quality and hormones with varying results. Bonde and Ernst[141] did not find correlations between urinary chromium and semen parameters in a study of mild steel welders, nonwelding metal workers, and electricians, where the majority of urinary chromium levels were low across the study groups. In contrast, both Li et al.[135] and Chen et al.[142] reported reduced sperm count and motility in workers at chromium-electroplating factories. The investigation by Li et al.[135] illustrates some of the most common limitations encountered when evaluating epidemiologic work with respect to metals and male reproduction. The researchers divided men in an electroplating factory into 21 exposed and 22 unexposed workers, based on reputed exposure to Cr(VI) using a definition (not delineated in the publication) of exposure to harmful chemicals. Statistically significant reductions in sperm count, motility, zinc, LDH, and LDH-C_4 were found in semen of the exposed group ($p < 0.05$) as well as increased FSH in blood ($p < 0.01$). This profile is consistent clinically with reduced fertility in human males. The researchers did not report fertility history in the study population, probably because numbers of subjects were not adequate to be informative (approximately 500 workers would be needed for this parameter in order to identify an OR of 2.5 with 80% power). Control for nutrition and smoking was not mentioned. Men grouped on exposure, based on reputed exposure to Cr(VI), did not differ significantly on measures of elemental chromium in serum (Cr in $\mu mol/ml$ $1.40 \pm 0.01 \times 10^{-3}$ exposed, $1.26 \pm 0.02 \times 10^{-3}$ controls) or seminal fluid (Cr in $\mu mol/ml$ $7.55 \pm 0.06 \times 10^{-3}$ exposed, $6.38 \pm 1.06 \times 10^{-3}$ controls), although only 4 of 22 controls were assayed for seminal fluid chromium. Chromium species was not delineated. The finding of no difference in exposed and unexposed groups in serum or seminal fluid chromium could suggest: low exposures, misclassification of exposed and unexposed, failure to adequately determine the correct metal species related to exposure,[143] or contribution of the metal of interest through nonwork exposures such as smoking or nutrition. This allows for an entirely different factor, other than Cr(VI), to be the cause of the poor human semen quality detected in the population. Many early epidemiologic investigations of metals fall prey to misclassification, lack of detailed exposure information, failure to report metal species, inadequate control for confounders, and inadequate power. The finding of no statistically significant difference in the levels of metals of interest in biological fluids of men grouped as exposed or unexposed based on job tasks has been reported in several recent epidemiologic studies conducted in developed countries.[55,135] This could reflect improved industrial hygiene, but highlights that misclassification can occur when relying on job title exclusively without biological markers of exposure.

Magnesium

Of all the metals listed in Table 8.2, magnesium appears to be the least toxic to human sperm at the levels investigated. Magnesium is present in seminal fluid at concentrations approximately three times that of blood. It is highly correlated with seminal fluid Zn ($r = 0.73$, $p < 0.0001$), a critical element in male reproduction.[53] Pandy et al.[59] reported low seminal fluid Mg levels in subfertile men although Wong et al.[53] found seminal fluid levels of magnesium were not statistically different between infertile and control men (mean concentration of 3.6 ± 1.1 mM Mg in 103 subfertile males versus 3.7 ± 1.0 mM Mg in 107 fertile males), but were significantly correlated with sperm count (r = 0.16, p < 0.05) when combining the subfertile and fertile groups. In a study investigating time to pregnancy in 50 couples related to seminal fluid levels of zinc and magnesium, Sorensen et al.[48] also found a strong correlation between seminal fluid zinc and magnesium (r = 0.86, p < 0.01) but found no correlation with semen parameters (median magnesium level in seminal fluid of males achieving pregnancy in less than two months 86 mg/l and males taking over 7 months 100 mg/l). This suggests magnesium and zinc are related, although the exact function of the relationship is not clear. Three additional studies found no significant correlation between magnesium levels in seminal plasma and fertility status.[43,46,47]

Manganese

Manganese, the fourth most utilized metal in terms of tonnage, is found in metallurgy, dry-cell batteries, glass, ceramics, dyes, pigments, pesticides, gasoline, and medicinals.[121] It is an essential trace element. Similar to magnesium, zinc, and calcium ions, manganese *in vitro* has been shown to stimulate progressive motility of ejaculated sperm in a time-dose-dependent manner.[144] In an *in vitro* investigation of manganese-induced lipid peroxidation of sperm membrane measured as malondialdehyde (MDA), Huang et al.[145] failed to find any significant increase in MDA levels even at the highest exposure level of 500 ppm. An occupational epidemiologic study by Lauwerys et al.[30] found a statistically significant difference in the expected vs. observed number of children fathered in 81 controls (blood manganese levels 0.04 to 1.31 µg/dl, mean 0.57 µg/dl) vs. 85 manganese-exposed workers recruited from a factory producing manganese salts from concentrated ores (blood manganese levels ranging from 0.10 to 3.30 µg/dl, mean 1.29 µg/dl). Gennart et al.[31] found no significant difference in probability of live births to 70 workers exposed to manganese dioxide (median atmospheric concentration of total manganese dust 0.71 mg/m³) in a dry alkaline battery plant compared to 138 controls. Hjollund et al.[61,62] found no difference in semen quality, sex hormones, or fertility measured as time to pregnancy in welders putatively exposed to manganese through welding fumes; however, there were no differences in urine concentrations of manganese between welders and controls, or within welders pre- and postwork shifts suggesting exposures were well below an observable adverse effect level. In a study comparing infertile men, Abou-Shakra et al.[47] reported seminal plasma values of manganese in 19 normospermic men that were lower than 19 oligospermic men (0.017 ± 0.006 µg/ml versus 0.024 ± 0.008), but argue several places in the paper that seminal plasma may not be the most appropriate fluid to use because of homeostatic regulatory mechanisms.

Mercury

Mercury is a ubiquitous environmental contaminant shown to have effects on male reproduction in animal studies and other model test systems (discussed in a previous chapter of this text). Consumption of contaminated seafood is a significant current source of exposure in some human populations. *In vitro* work demonstrates binding of mercury to sulfhydryl groups in human sperm membrane, head, midpiece, and tail.[146] Two occupational epidemiologic studies looking at spontaneous abortion found a dose response in Relative Risk (RR) for spontaneous abortion and/or stillbirth with increasing urinary levels of mercury in fathers. The Alcser et al.[40] study of exposure to elemental mercury was based on paternal report of pregnancy loss in relationship to company records and biological monitoring in a U.S. energy plant. A 7% increase in risk for each increment of 100 µg/l mercury in urine was found. The Cordier et al.[41] data was based on maternal report of pregnancy loss in relationship to company records of exposure to mercury vapor and biological monitoring in a chloralkali plant in France, and reported increasing adjusted odds of 1.3 (95% CI = 1.0 to 1.7), 1.7 (95% CI = 1.0 to 3.0), and 2.3 (95% CI = 1.0 to 5.2) for urinary mercury levels of 1 to 19 µg/l, 20-49 µg/l, and greater than 49 µg/l, respectively. A third occupational epidemiologic study[30] with lower exposure levels, found no statistically significant difference in the expected vs. observed number of children fathered in 101 controls (blood mercury levels 0.06 to 0.56 µg/dl, mean 0.21µg/dl) vs. 103 mercury-exposed workers from a zinc-mercury amalgam factory, chloralkali plant, and electrical equipment manufacturing plant (blood mercury levels ranging from 0.25 to 6.42 µg/dl, mean 1.46 µg/dl). And a fourth occupational study[36] ascertaining exposure through self-report on a mailed questionnaire did not find an increased risk for spontaneous abortion in dentists putatively exposed during preparation of mercury amalgams (RR 0.9, 95% CI 0.7 to 1.1). One occupational case report showed effects on sexual performance at a single acute exposure to mercury vapor.[147] In the andrology literature, two studies investigated blood mercury and semen parameters. In a clinic population of 111 subfertile men undergoing IVF, Choy et al.[50] measured total mercury in blood (average 41.4 ± 1.7 nmol/l) and seminal fluid (average 22.1 ± 2.0 nmol/l) and found statistically significant correlations between seminal fluid mercury concentrations and sperm head and midpiece defects plus adverse effects on the motion characteristics of VSL (straight line velocity), LIN (linearity), ALH (lateral head displacement), and VAP (average path velocity). Chia et al.,[57] using a Singapore infertility clinic population of 34 subjects, found no associations between blood mercury levels (mean 18.4±6.6 µg/dl) and conventional semen parameters. Although it is difficult to reconcile the exposure levels across these mercury studies because of lack of reporting of metal species and different biological matrices used, it appears there could be dose-dependent effects on seminal parameters that might predict adverse pregnancy outcomes.

Because risk of exposure to mercury through food in some human populations in still a concern, future work on exposure levels that cause male reproductive effects is warranted.

Nickel

Workplace exposure to nickel can occur in electroplating, welding, flame cutting, mold making, jewelry, coinage, and the manufacture of cutlery, cooking utensils, and dental or surgical prostheses.[148] Nickel is a known human carcinogen and weak mutagen with epigenetic effects.[149,150] There are only a few human studies that have been published looking specifically at nickel as a male reproductive toxicant. The only study that was adequately powered based on highly exposed workers (average blood nickel levels of 16.7±5.8 µg/l controls and 123.3±35.2 µg/l exposed workers) was conducted in southern India and reported by Danadevi et al.[58] Among the exposed workers, statistically significant correlations were found between blood nickel concentrations and the semen parameters of rapid linear progressive motility ($r = -0.38$, $p = 0.045$), slow/nonlinear progressive motility ($r = 0.39$, $p = 0.042$), tail defects ($r = 0.48$, $p = 0.036$), and sperm vitality ($r = -0.42$, $p = 0.026$). In a study of time to pregnancy with low exposures (average blood nickel concentrations of 0.02 to 0.35 µg/dl), Figa-Talamanca et al.[55] failed to demonstrate statistically significant increased time to pregnancy in 167 nickel-exposed metal workers, although findings suggested time to pregnancy was increased in the exposed mint workers. In the Figa-Talamanca et al.[55] study, reproductive history and sociodemographic information was collected using a standardized personal interview schedule,[151] and was used to calculate time to pregnancy across groups considered to be exposed and unexposed. Interestingly, a significant decrease in male births was recorded for the most recent 43 births of the men exposed to nickel and chromium through foundry work (30.2% males, $p = 0.04$). Because gender ratios at birth are a sensitive indicator of reproductive toxicity, this finding suggests nickel may affect the male reproductive system and further work is necessary.[56] Hjollund et al.[61,62] found no difference in semen quality, sex hormones, or fertility measured as time to pregnancy in welders putatively exposed to nickel through welding fumes, but there were no differences in urine concentrations of nickel between welders and controls, or within welders' pre- and postwork shifts, suggesting exposures were low.

Welding

Welding fume particulates contain metals such as manganese, hexavalent chromium, and nickel. Monitoring of urine and blood of welders and workers exposed to welding has confirmed biological uptake of metals.[71] Because approximately 1% of the workforce is estimated to be exposed to welding, a number of epidemiological studies have looked at this job exposure for evidence of effects of metals on male reproductive outcomes (Table 8.4).

Different welding processes confer different metal exposure risks, with chromium and nickel associated primarily with stainless steel welding.[71] Stainless steel is an alloy of iron, nickel, and chromium, occasionally containing cobalt, vanadium, manganese, and molybdenum. Mild steel is an alloy of iron, carbon, and silicon, occasionally containing molybdenum or manganese. Welding also involves exposures to heat, electromagnetic fields, and organic solvents that could adversely affect the male reproductive system, making it difficult to attribute adverse reproductive effects only to metals.

Lindbohm et al.[152] reported sperm counts less than 4×10^6 spermatozoa/ml (below WHO guidelines for normal sperm counts) in more than 50% of men in an investigation of 61 welders in Finland. In a multicenter case-control study, Mortensen et al.[153] reported reduced sperm quality among metal workers with special reference to 55 welders. Levels of chromium in seminal fluid were elevated in welders compared with nonwelders. Following this, Bonde and his research group conducted a comprehensive evaluation of effects of welding on male reproductive health through a series of investigations starting with a cohort of Danish male welders assembled between 1964 and 1984. Bonde et al.[154] reported reduced fertility in the cohort during exposed years for mild steel welders but not stainless steel welders. Impaired fertility was not associated with duration of exposure, suggesting the exposures were not cumulative.[61,154] Later, Bonde and Ernst[141] reported poor semen quality and reduced fertility in a subset of the Danish welders. Bigelow et al.[119] reported a significantly higher percent of coiled tail defects and decreased progressive motility in welders compared to controls. In a more recent study conducted in India, Danadevi et al.[58] compared welders exposed to fumes for 2 to 21 years with control subjects matched for age, lifestyle, and economic status. Conventional semen parameters of count, motility, normal morphology, vitality and nonspecific agglutination were poorer for the welders (p < 0.001). In addition, they modeled nickel and chromium levels in the blood with semen parameters of a subset of 28 welders and 27 controls. Although welders as a group had significantly poorer sperm count, motility, morphology, and vitality, only tail defects (r = 0.485, p = 0.036), vitality (r = –0.42, p = 0.026), and progressive motility (r = –0.38, p = 0.045) were significantly associated with blood nickel, whereas count (r= –0.424, p = 0.025) as well as progressive motility (r = –0.48, p = 0.009), tail defects (r = 0.485, p = 0.009), and vitality (r = –0.507, p = 0.006) were associated with blood chromium. Of note, tail defects were associated with blood levels of nickel and chromium in both controls and welders, but an overall correlation coefficient for the metals was not reported.

Not all studies report significant effects. Jelnes et al.[155] did not find effects on semen quality associated with stainless steel welding. Hjollund et al.[61] conducted a well-designed population-based study of welders that enrolled primiparous couples when they discontinued birth control in order to achieve a pregnancy. Detailed information on type of welding, duration of exposure, engineering protection, and urinary concentrations of chromium, manganese, and nickel were collected and correlated with semen parameters including computer-assisted motility indices. Information on exposures to other potential toxins, smoking, age, and urogenital health was also collected. Based on a 16% participation rate resulting in 394 participants, no statistically significant differences were found between welders, nonwelders, and nonmetal workers. However, no differences were noted in urinary concentrations of chromium, manganese, or nickel between the exposure groups or between preshift and postshift measures, suggesting low exposures explain the lack of significance for welding in general, and chromium, manganese, and nickel specifically.

Looking at the body of published work on welding, it appears there is an association between this occupation and adverse male reproductive outcomes. It is not clear if the effect on male reproduction is an interaction between physical hazards associated with the work, for example heat, or if the association is related to combinations of metals, or is due to other chemicals used in the work processes such as solvents. It is also evident that metal exposures associated with welding seem to be decreasing in the present workforce when comparing more recent studies to past studies conducted in Denmark, possibly due to improved industrial hygiene and changes in the work process. This may explain some of the variability in effect measures detected and reported during the accumulation of welding-based data over time.

Summary

A moderate number of epidemiologic studies and subsequent review papers have generated and evaluated substantial data on human male reproductive toxicity related to exogenous metal exposures. Integrating principles are as follows. First, synthesis is impracticable across the epidemiologic research because consistency in reporting of exposures across studies is lacking, even when biomarkers of exposure are utilized (e.g., metal species not reported, some reported urinary levels are corrected for creatinine whereas others are not, exposure timing in relation to collection of biological specimens is not reported, assays have changed over time). Second, occupational cohorts assembled to investigate male reproductive effects from metals within the past 15 years in the United States and Western Europe report far less metal exposure than prior worker cohorts. Current workplace conditions likely reflect adherence to recommended action levels for metals proposed by governmental and professional agencies (i.e., World Health Organization, National Institute for Occupational Safety and Health, American Conference of Industrial Hygienists). Consequently, traditional occupational epidemiologic approaches are not likely to detect effects as a result of prevailing U.S. workplace exposures, and thus measures of effects reported more recently often conflict with previous reports. Third, there have been improvements in outcome determinations over the course of the body of work. For example, statistical approaches for fecundity measures are more sophisticated, computer-assisted semen analysis contributes a new layer of sensitivity to semen motility measures ascertained in the field, and so forth. This also adds to the variability seen in effect measures reported in the body of work over time. In the long run, however, as refined assays and statistical techniques are implemented, improvements in ascertainment of important associations between metals and male reproductive health will emerge.

Fourth, there are increasingly more publications from andrological clinical settings on metal exposures specifically, and environmental exposures in general. Although these populations have specific limitations, for example, relatively lower metal exposures compared to past occupational cohorts, small differences between exposed and unexposed groups, lack of precision in source of exposure, single metal assessments, and limited generalizability, it appears the ease of biologic specimen collection and sensitivity of biologic and medical markers provide improved ability to detect associations and generate valuable information. Thus, Wong et al.[53] were able to report seminal fluid magnesium levels that demonstrate significant positive correlations with sperm count across fertile and subfertile populations, Choy et al.[50] were able to report on seminal fluid mercury levels that correlate with sperm concentration, morphology, motion characteristics; and Chia et al.[32,57] were able to report blood and seminal fluid cadmium levels that correlate with sperm density and morphology as well as blood mercury levels that show no effect on conventional semen parameters. This venue for human field studies remains important and should continue to be considered in future research.

Finally, in spite of study limitations and inconsistencies in effects reported across the epidemiologic studies over time, there are some trends in the accumulated data. Lead and welding have clearly been shown to affect aspects of male reproduction. Boron has repeatedly been shown to influence the expected male-to-female birth ratios in the epidemiologic investigations that looked at this metalloid and birth gender ratio. Mercury and cadmium are suspect and warrant further investigation. Overall, study designs need to be improved to allow ascertainment of simultaneous metal exposures and interaction of metals with physical and other chemical agents. In addition, newer contemporary assays of exposure and outcome need to be incorporated into future designs. Studies conducted in andrological clinic populations should continue to be pursued because of the availability of sensitive and specific outcome measures plus comprehensive data on confounders. Future research utilizing all of these techniques can allow better ascertainment of important associations between exogenous metal exposures and effects on human male reproductive health.

References

1. Carlsen, E. et al., Evidence for decreasing quality of semen during the past 50 years, *Br. Med. J.*, 305, 609, 1992.
2. Leke, R.J. et al., Regional and geographical variations in infertility: effects of environmental, cultural, and socioeconomic factors, *Environ. Health Perspect.*, 101 (Suppl. 2), 73, 1993.

3. Giwercman, A. et al., Evidence for increasing incidence of abnormalities of the human testis: a review, *Environ. Health Perspect.*, 101 (Suppl. 2), 65, 1993.
4. Thompson, S.T., Prevention of male infertility: an update, *Urologic Clinics of North Am.*, 21, 365, 1994.
5. Auger, J. et al., Decline in semen quality among fertile men in Paris during the past 20 years, *N. Engl. J. Med.*, 332, 281, 1995.
6. Toppari, J. et al., Male reproductive health and environmental xenoestrogens, *Environ. Health Perspect.*, 104, (Suppl. 4), 741, 1996.
7. Auger, J. and Jouannet, P., Evidence for regional differences of semen quality among fertile French men. Federation Française des Centres d'Etude et de Conservation des Oeufs et du Sperme humains, *Hum. Reprod.*, 12, 740, 1997.
8. Swan, S.H., Elkin, E.P., and Fenster, L., The question of declining sperm density revisited: an analysis of 101 studies published 1934-1996, *Environ. Health Perspect.*, 108, 961, 2000.
9. Jorgensen, N. et al., Regional differences in semen quality in Europe, *Hum. Reprod.*, 16, 1012, 2001.
10. *International Union v. Johnson Controls*, 111 U.S. 1196, 1991.
11. Anttila, A. and Sallmen, M., Effects of parental occupational exposure to lead and other metals on spontaneous abortion, *J. Occup. Environ. Med.*, 37, 915, 1995.
12. Baranski, B., Effects of the workplace on fertility and related reproductive outcomes, *Environ. Health Perspect.*, 101 (Suppl. 2), 81, 1993.
13. Figa-Talamanca, I., Traina, M.E., and Urbani, E., Occupational exposures to metals, solvents and pesticides: recent evidence on male reproductive effects and biological markers, *Occup. Med.*, 51, 174, 2001.
14. Lahdetie, J., Occupation and exposure related studies on human sperm, *J. Occup. Environ. Med.*, 37, 922, 1995.
15. Lamb, E.J. and Bennett, S., Epidemiologic studies of male factors in infertility, *Ann. NY Acad. Sci.*, 709, 165, 1994.
16. Narod, S.A. et al., Human mutagens: evidence from paternal exposure? *Environ. Mol. Mutagen.*, 11, 401, 1988.
17. Olsen, J.H. et al., Parental employment at time of conception and risk of cancer in offspring, *Eur. J. Cancer*, 27 (8), 958, 1991.
18. Olshan, A.F., Teschke, K., and Baird, P.A., Paternal occupation and congenital anomalies in offspring, *Am. J. Ind. Med.*, 20, 447, 1991.
19. Robbins, W.A. and Cousins, D.S., Reproductive hazards in the workplace, *Infertil. Reprod. Med. Clin. North Am.*, 9, 545, 1998.
20. Rosenberg, M.J., Feldblum, P.J., and Marshall, E.G., Occupational influences on reproduction: a review of recent literature, *J. Occup. Med.*, 29, 584, 1987.
21. Savitz, D.A. and Chen, J., Parental occupation and childhood cancer, review of epidemiological studies, *Environ. Health Perspect.*, 88, 325, 1990.
22. Savitz, D.A., Sonnenfiled, L.N., and Olshan, F.A., Review of epidemiologic studies of paternal occupational exposure and spontaneous abortion, *Am. J. Ind. Med.*, 25, 361, 1994.
23. Sheiner, E.K. et al., Effect of occupational exposures on male fertility: literature review, *Ind. Health*, 41, 55, 2003.
24. Steeno, O.P. and Pangkahila, A., Occupational influences on male fertility and sexuality, *Andrologia*, 16, 5, 1983.
25. Tas, S., Lauwerys, R., and Lison, D., Occupational hazards for the male reproductive system, *Crit. Rev. Toxicol.*, 26, 261, 1996.

26. Tarasenko, N.Y., Kasparov, A.A., and Strongina, O.M., Effect of boric acid on the generative function in male, *Gigiena Truda Professionalnye Zabolevaniya*, 11, 13, 1972.
27. Whorton, M.D. et al., Reproductive effects of sodium borates on male employees: birth rate assessment, *Occup. Environ. Med.*, 51, 761, 1994.
28. Sayli, B.S., Low frequency of infertility among workers in a borate processing facility, *Biol. Trace Element Res.*, 93 (1–3), 19, 2003.
29. Robbins, W.A. et al., Reproductive health of male boron workers in NE Region of PR China, presented at American Society of Andrology, Baltimore, MD, 2004.
30. Lauwreys, R. et al., Fertility of male workers exposed to mercury vapor or to manganese dust: a questionnaire study, *Am. J. Ind. Med.*, 7, 171, 1985.
31. Gennart, J.P. et al., Fertility of male workers exposed to cadmium, lead, or manganese, *Am. J. Epidemiol.*, 135, 1208, 1992.
32. Chia, S.E. et al., Effect of cadmium and cigarette smoking on human semen quality, *Int. J. Fertil.*, 39, 292, 1994.
33. Keck, C. et al., Lack of correlation between cadmium in seminal plasma and fertility status of nonexposed individuals and two cadmium-exposed patients, *Reprod. Toxicol.*, 9 (1), 35, 1995.
34. Saaranen, M. et al., Human seminal plasma cadmium: comparison with fertility and smoking habits, *Andrologia*, 21, 140, 1989.
35. Mason, H.J., Occupational cadmium exposure and testicular endocrine function, *Hum. Exp. Toxicol.*, 9 (2), 91, 1990.
36. Brodsky, J.B. et al., Occupational exposure to mercury in dentistry and pregnancy outcome, *J. Am. Dent. Assoc.*, 11, 770, 1985.
37. Lindbaum, M.L. et al., Paternal occupational lead exposure and spontaneous abortion, *Scand. J. Work Environ. Health*, 17, 95, 1991.
38. Popescu, H.I., Poisoning with alkylmercury compounds, *Br. Med. J.*, 1 (6123), 1347, 1978.
39. Barregard, L. et al., Endocrine function in mercury exposed chloralkali workers, *Occup. Environ. Med.*, 51, 536, 1994.
40. Alcser, K.H. et al., Occupational mercury exposure and male reproductive health, *Am. J. Ind. Med.*, 15, 517, 1989.
41. Cordier, S. et al., Paternal exposure to mercury and spontaneous abortions, *Br. J. Ind. Med.*, 48, 375, 1991.
42. Emara, A.M. et al., Chronic manganese poisoning in the dry battery industry, *Br. J. Ind. Med.*, 28, 78, 1971.
43. Umeyama, T. et al., A comparative study of seminal trace elements in fertile and infertile men, *Fertil. Steril.*, 46, 494, 1986.
44. Xu, G. et al., Trace elements in blood and seminal plasma and their relationship to sperm quality, *Reprod. Toxicol.*, 7, 613, 1993.
45. Xu, D.-X. et al., The associations among semen quality, oxidative DNA damage in human spermatozoa and concentrations of cadmium, lead and selenium in seminal plasma, *Mut. Res.*, 534, 155, 2003.
46. Saaranen, M. et al., Lead, magnesium, selenium and zinc in human seminal fluid: comparison with semen parameters and fertility, *Hum. Reprod.*, 2, 475, 1987.
47. Abou-Shakra, F.R., Ward, N.I., and Everard, D.M., The role of trace elements in male infertility, *Fertil. Steril.*, 52, 307, 1989.
48. Sorensen, M.B. et al., Zinc, magnesium and calcium in human seminal fluids: Relations to other semen parameters and fertility, *Mol. Hum. Reprod.*, 5, 331, 1999.

49. Oldereid, N.B., Thomassen, Y., and Purvis, K., Selenium in human male reproductive organs, *Hum. Reprod.*, 13, 2172, 1998.

50. Choy, C. et al., Relationship between semen parameters and mercury concentrations in blood and in seminal fluid from subfertile males in Hong Kong, *Fertil. Steril.*, 78, 426, 2002.

51. Adejuwon, C.A. et al., Biophysical and biochemical analysis of semen in infertile Nigerian males, *Afr. J. Med. Sci.*, 25, 217, 1996.

52. Scott, R. et al., The effect of oral selenium supplementation on human sperm motility, *Br. J. Urol.*, 82, 76, 1998.

53. Wong, W.Y. et al., The impact of calcium, magnesium, zinc, and copper in blood and seminal plasma on semen parameters in men, *Reprod. Toxicol.*, 15, 131, 2001.

54. Telisman, S. et al., Semen quality and reproductive endocrine function in relation to biomarkers of lead, cadmium, zinc, and copper in men, *Environ. Health Perspect.*, 108, 45, 2000.

55. Figa-Talamanca, I. et al., Fertility of male workers of the Italian mint, *Reprod. Toxicol.*, 14, 325, 2000.

56. Figa-Talamanca, I. and Petrelli, G., Reduction in male births among workers exposed to metal fumes, Letters to the Editor, *Int. J. Epidemiol.*, 29, 381, 2000.

57. Chia, S.E. et al., Blood concentrations of lead, cadmium, mercury, zinc, and copper and human semen parameters, *Arch. Androl.*, 29, 177, 1992.

58. Danadevi, K. et al., Semen quality of Indian welders occupationally exposed to nickel and chromium, *Reprod. Toxicol.*, 17, 451, 2003.

59. Pandy, V.K., Parmeshwaran, M., and Soman, S.D., Concentrations of morphologically normal, motile spermatozoa: Mg, Ca and Zn in the semen of infertile men, *The Science of the Total Environ.*, 27, 49, 1983.

60. Hovatta, O. et al., Aluminum, lead, and cadmium concentrations in seminal plasma and spermatozoa, and semen quality in Finnish men, *Hum. Reprod.*, 13, 115, 1998.

61. Hjollund, N.H.I. et al., A follow-up study of male exposure to welding and time to pregnancy, *Reprod. Toxicol.*, 12, 29, 1998.

62. Hjollund, N.H.I. et al., Semen quality and sex hormones with reference to metal welding, *Reprod. Toxicol.*, 12, 91, 1998.

63. Sayli, B.S., An assessment of fertility in boron exposed Turkish subpopulations, *Biol. Trace Elem. Res.*, 66, 409, 1998.

64. Siemons, L.J. and Mahler, C.H., Hypogonadotropic hypogonadism in hemochromatosis: recovery of reproductive function after iron depletion, *J. Clin. Endocrinol Metab.*, 65, 585, 1987.

65. De Sanctis, V., Growth and puberty and its management in thalassaemia, *Horm. Res.*, 58 (Suppl. 1), 72, 2002.

66. Keskes-Ammar, L. et al., Sperm oxidative stress and the effect of an oral vitamin E and selenium supplement on semen quality in fertile men, *Arch. Androl.*, 49, 83, 2003.

67. Bleau, G. et al., Semen selenium and human fertility, *Fertil. Steril.*, 42, 890, 1984.

68. Taskinen, H.K., Epidemiological studies in monitoring reproductive effects, *Environ. Health Perspect.*, 101 (Suppl. 3), 279, 1993.

69. Carter, D.E., Oxidation-reduction reactions of metal ions, *Environ. Health Perspect.*, 103 (Suppl. 1), 17, 1995.

70. Savitz, D.A., Poole, C., and Miller, W.C., Reassessing the role of epidemiology in public health, *Am. J. Public Health*, 89, 1158, 1999.

71. Bonde, J.P., The risk of male subfecundity attributable to welding of metals, *Int. J. Androl.*, 16 (Suppl. 1), 1, 1993.
72. *WHO Laboratory Manual for the Examination of Human Semen and Sperm-Cervical Mucus Interaction*, 4th ed., Cambridge University Press, New York, 1999, chap. 1–4.
73. Jounannet, P. et al., Male factors and the likelihood of pregnancy in infertile couples. I. Study of sperm characteristics, *Int. J. Androl.*, 11, 379, 1988.
74. Bonde, J.P. et al., Relation between semen quality and fertility: a population-based study of 430 first-pregnancy planners, *Lancet*, 352, 1172, 1998.
75. Horvath, P.M. et al., The relationship of sperm parameters to cycle fecundity in superovulated women undergoing intrauterine insemination, *Fertil. Steril.*, 52, 288, 1989.
76. Chia, S.E. et al., Factors associated with male infertility a case-control study of 218 infertile and 240 fertile men, *Br. J. Obstet. Gynaecol.*, 107, 55, 2000.
77. Perreault, S.D. et al., Integrating new tests of sperm genetic integrity into semen analysis: breakout group discussion, *Adv. Exp. Med. Biol.*, 518, 253, 2003.
78. Richthoff, J. et al., The impact of testicular and accessory sex gland function on sperm chromatin integrity as assessed by the sperm chromatin structure assay, *Hum. Reprod.*, 17, 3162, 2002.
79. Abma, J.C. et al., Fertility, family planning, and women's health: new data from the 1995 National Survey of Family Growth, Hyattsville, MD: U.S. Department of Health and Human Services, Centers for Disease Control, National Center for Health Statistics, Vital and Health Statistics, 23, 1997.
80. Williams, M. et al., Subfertility and the risk of low birth weight, *Fertil. Steril.*, 56, 688, 1991.
81. Joffe, M. and Li, Z., Association of time to pregnancy and the outcome of pregnancy, *Fertil. Steril.*, 62, 71, 1994.
82. Henriksen, T.B. et al., Time to pregnancy and preterm delivery, *Obstet. Gynecol.*, 89, 594, 1997.
83. Joffe, M., Invited commentary: the potential for monitoring of fecundity and the remaining challenges, *Am. J. Epidemiol.*, 157, 89, 2003.
84. Tingen, C. et al., Methodologic and statistical issues in studying human fertility and its relationship to reproductive, perinatal, and child health, *Environ. Health Perspect.*, 112, 87, 2002.
85. Whorton, D., Haas, J.L., and Trent L., Reproductive effects of inorganic borates on male employees: birth rate assessment, *Environ. Health Perspect.*, 102 (suppl. 7), 129, 1994b.
86. Regal, R.R. and Hook, E.B., Interrelationships of relative risks of birth defects in embryonic and fetal deaths, in live births, and in all conceptuses, *Epidemiology*, 3, 247, 1992.
87. Wilcox, A.J. and Horney, F.L., Accuracy of spontaneous abortion recall, *Am. J. Epidemiol.*, 120, 727, 1984.
88. *Guidelines for Studies of Human Populations Exposed to Mutagenic and Reproductive Hazards*, Bloom, A.D., ed., March of Dimes Birth Defects Foundation, White Plains, NY, 1981, 47.
89. Palmer, A.K., Identifying environmental factors harmful to reproduction, *Environ. Health Perspect.*, 101 (Suppl. 2), 19, 1993.
90. Paul, M., Occupational reproductive hazards, *Lancet*, 349, 1385, 1997.
91. Nelson, K. and Holmes, L.B., Malformations due to presumed spontaneous mutations in newborn infants, *N. Engl. J. Med.*, 320, 19, 1989.

92. Bonde, J.P., Semen quality in welders exposed to radiant heat, *Br. J. Ind. Med.*, 49, 5, 1992.
93. Buckley, J.D. et al., A case-control study of risk factors for hepatoblastoma, *Cancer*, 64, 1169, 1989.
94. Wilkins, J.R. and Koutras, R.A., Paternal occupation and brain cancer in offspring: a mortality-based case-control study, *Am. J. Ind. Med.*, 14, 299, 1988.
95. Heller, C.G. and Clermont, Y., Kinetics of the germinal epithelium in man, *Recent Prog. Horm. Res.*, 20, 545, 1964.
96. Sharpe, R.M., Regulation of spermatogenesis, in *The Physiology of Reproduction*, 2nd ed, Knobil, E. and Neill, J.D., Eds., Raven Press, New York, 1994, 1363–1434.
97. Chapin, R.E. et al., History matters: the influence of pre- and peri-conceptional exposures, parental fertility, and nutrition on childrens' health, accepted, *Environ. Health Perspect.*, 112, 69, 2003.
98. Field, A.E. et al., The relation of smoking, age, relative weight, and dietary intake to serum adrenal steroids, sex hormones, and sex hormone-binding globulin in middle-aged men, *J. Clin. Endocrinol. Metab.*, 79, 1310, 1994.
99. Vine, M.F. et al., Cigarette smoking and sperm density: a meta-analysis, *Fertil. Steril.*, 61, 35, 1994.
100. Vine, M.F., Smoking and male reproduction: a review, *Int. J. Androl.*, 19, 323, 1996.
101. Curtis, K.M., Savitz, D.A., and Arbuckle, T.E., Effects of cigarette smoking, caffeine consumption, and alcohol intake on fecundability, *Am. J. Epidemiol.*, 146, 32, 1997.
102. Robbins, W.A. et al., Use of FISH (fluorescence in situ hybridization) to assess effects of smoking, caffeine, and alcohol on aneuploidy load in sperm of healthy men, *Environ. Mol. Mutagen.*, 30, 175, 1997.
103. Mak, V. et al., Smoking is associated with the retention of cytoplasm by human spermatozoa, *Urology*, 56, 463, 2000.
104. Wong, W.Y. et al., Cigarette smoking and the risk of male factor subfertility: minor association between cotinine inseminal plasma and semen morphology, *Fertil. Steril.*, 74, 930, 2000.
105. Olsen, J. et al., Tobacco use, alcohol consumption and infertility, *Int. J. Epidemiol.*, 12, 179, 1983.
106. Florack, E.M., Zielhous, G.A., and Rolland, R., Cigarette smoking, alcohol consumption and caffeine intake and fecundability, *Prev. Med.*, 23, 175, 1994.
107. Goverde, H.J.M. et al., Semen quality and frequency of smoking and alcohol consumption-an explorative study, *Int. J. Fertil.*, 40, 135, 1995.
108. Harkonen, K. et al., Aneuploidy in sperm and exposure to fungicides and lifestyle factors, ASCLEPIOS A European Concerted Action on Occupational Hazards to Male Reproductive Capability, *Environ. Mol. Mutagen.*, 34, 39, 1999.
109. Jordan, N.P. et al., Effects of caffeine on human health, *Food Addit. Contam.*, 20, 1, 2003.
110. Thonneau, P. et al., Occupational heat exposure and male fertility: a review, *Hum. Reprod.*, 13, 2122, 1998.
111. Bujan, L. et al., Increase in scrotal temperature in car drivers, *Hum. Reprod.*, 15, 1355, 2000.
112. Levine, R.J. et al., Air-conditioned environments do not prevent deterioration of human semen quality during the summer, *Fertil. Steril.*, 57, 1075, 1992.

113. Auger, J. et al., Sperm morphological defects related to environment, lifestyle and medical history of 1001 male partners of pregnant women from four European cities, *Hum. Reprod.*, 16, 2710, 2001.
114. Henkel, R. et al., Seasonal changes in human sperm chromatin condensation, *J. Assist. Reprod. Genet.*, 18, 371, 2001.
115. Krause, A. and Krause, W., Seasonal variations in human seminal parameters, *Eur. J. Obstet. Gynaecol. Reprod. Biol.*, 101, 175, 2002.
116. Chen, Z. et al., Seasonal variation and age-related changes in human semen parameters, *J. Androl.*, 24, 226, 2003.
117. Robbins, W.A. et al., Chemotherapy induces transient sex chromosomal and autosomal aneuploidy in human sperm, *Nature Genet.*, 16, 74, 1997.
118. Wyrobek, A.J. et al., Assessment of reproductive disorders and birth defects in communities near hazardous chemical sites III. Guidelines for field studies of male reproductive disorders, *Reprod. Toxicol.*, 11, 243, 1997.
119. Bigelow, P.L. et al., Association of semen quality and occupational factors: comparison of case-control analysis and analysis of continuous variables, *Fertil. Steril.*, 69, 11, 1998.
120. Evenson, D.P. et al., Characteristics of human sperm chromatin structure following an episode of influenza and high fever: a case study, *J. Androl.*, 21, 739, 2000.
121. Gerber, G.B., Leonard, A., and Hantson, P.H., Carcinogenicity, mutagenicity and teratogenicity of manganese compounds, *Crit. Rev. Oncol. Hematol.*, 42, 25, 2002.
122. Riffo, M., Leiva, S., and Astudillo, J., Effect of zinc on human sperm motility and the acrosome reaction, *Int. J. Androl.*, 15, 229, 1992.
123. Eggert-Kruse, W. et al., Are zinc levels in seminal plasma associated with seminal leukocytes and other determinants of semen quality? *Fertil. Steril.*, 77, 260, 2002.
124. Zalups, R.K. and Ahmad, S., Molecular handling of cadmium in transporting epithelia, *Toxicol. Appl. Pharmacol.*, 186, 163, 2003.
125. Lichtlen, P. and Schaffner, W., The metal transcription factor MTF-1: biological facts and medical implications, *Swiss Med. Wkly*, 131, 647, 2001.
126. Vogel, F., Clinical consequences of heterozygosity for autosomal-recessive diseases, *Clin.Genet.*, 25, 381, 1984.
127. Rothman, K.J. and Greenland, S., Causation and causal inference, in *Modern Epidemiology*, Rothman, K.J. and Greenland, S., eds., Lippincott-Raven Publishers, Philadelphia, 1998, chap. 2.
128. De Roos, A.J. et al., Parental occupational exposures to chemicals and incidence of neuroblastoma in offspring, *Am. J. Epidemiol.*, 154, 106, 2001.
129. Marcus, M., Silbergeld, E., and Mattison, D., Research needs working group, a reproductive hazards research agenda for the 1990s, *Environ. Health Perspect.*, 101 (Suppl. 2), 175, 1993.
130. Woods, W.G., An introduction to boron: history, sources, uses, and chemistry, *Environ. Health Perspect.*, 102 (Suppl. 7), 5, 1994.
131. Wegman, D.H. et al., Acute and chronic respiratory effects of sodium borate particulate exposures, *Environ. Health Perspect.*, 102 (Suppl. 7), 119, 1994.
132. Moorman, W.J. et al., Prioritization of NTP reproductive toxicants for field studies, *Reprod. Toxicol.*, 14, 293, 2000.
133. Battersby, S., Chandler, J.A., and Borton, M.S., Toxicity and uptake of heavy metals by human spermatozoa, *Fertil. Steril.*, 37, 230, 1982.

134. Pant, N. et al., (2003) Lead and cadmium concentration in the seminal plasma of men in the general population: correlation with sperm quality, *Reprod. Toxicol.*, 17 (4), 447, 2003.

135. Li, H. et al., Effect of Cr (VI) exposure on sperm quality: human and animal studies, *Ann. Occup. Hyg.*, 45, 505, 2001.

136. Bonde, J.P. and Christensen, J.M., Chromium in biological samples from low-level exposed stainless steel and mild steel welders, *Arch. Environ. Health*, 46, 225, 1991.

137. Hagmar, L. et al., Chromosomal aberrations in lymphocytes predict human cancer: a report from the European Study Group on Cytogenetic Biomarkers and Health, *Cancer Res.*, 58, 4117, 1998.

138. Fabia, J. and Thuy, T., Occupation of father at time of birth of children dying of malignant diseases, *Br. J. Prev. Soc. Med.*, 28, 98, 1974.

139. Lowengart, R.A., Peter, J.M., and Cicioni, C., Childhood leukemia and parents' occupational and home exposures, *J. Nat. Cancer Inst.*, 79, 39, 1987.

140. Tomatis, L., Transgeneration carcinogenesis, a review of the experimental and epidemiological evidence, *Jpn. J. Cancer Res.*, 44, 443, 1994.

141. Bonde, J.P. and Ernst, E., Sex hormones and semen quality in welders exposed to hexavalent chromium, *Hum. Exp. Toxicol.*, 11, 259, 1992.

142. Chen, L.H. et al., Effect of Cr (VI) exposure on sperm quality: human and animal studies, *Ann. Occup. Hyg.*, 45, 505, 2001.

143. Duffus, J.H., Effect of Cr (VI) on sperm quality, Letter to the Editor, *Am. Occup. Hyg.*, 46, 269, 2001.

144. Magnus, O. et al., Effects of manganese and other divalent cations on progressive motility of human sperm, *Arch. Androl.*, 24, 159, 1990.

145. Huang, Y.-L., Tseng, W.-C., and Lin, T.-H., In vitro effects of metal ions (Fe^{2+}, Mn^{2+}, Pb^{2+}) on sperm motility and ipid peroxidation, *J. Toxicol. Environ. Health*, Part A, 62, 259, 2001.

146. Mohamed, M.K. et al., Laser light-scattering study of the toxic effects of methylmercury on sperm motililty, *J. Andrology*, 7, 11, 1986.

147. McFarland, R.B. and Reigel, H., Chronic mercury poisoning from a single brief exposure, *J. Occup. Med.*, 20, 532, 1978.

148. Agency for Toxic Substances and Disease Registry (ATSDR), *Toxicological Profile for Nickel*, U.S. Department of Health and Human Services, Atlanta, GA, 1997.

149. Costa, M. and Klein, C.B., Nickel carcinogenesis, mutation, epigenetics or selection, *Environ. Health Perspect.*, 57, 438, 1997.

150. Costa, M. et al., Molecular mechanisms of nickel carcinogenesis: gene silencing by nickel delivery to the nucleus and gene activation/inactivation by nickel-induced cell signaling, *J. Environ. Monitoring*, 5, 222, 2003.

151. Bolumar, F. et al., Smoking reduced fecundity: a European multicenter study on infertility and subfertility, *Am. J. Epidemiol.*, 143, 578, 1996.

152. Lindbaum, M.L., Hemminki, K., and Kyyronen, P., Parental occupational exposure and spontaneous abortions in Finland, *Am. J. Epidemiol.*, 120, 370, 1984.

153. Mortensen, J.T., Risk for reduced sperm quality among metal workers with special reference to welders, *Scand. J. Work Environ. Health*, 14, 37, 1988.

154. Bonde, J.P., Hansen, K.S., and Levine, R.J., Fertility among Danish male welders, *Scand. J. Work Environ. Health*, 16, 315, 1990.

155. Jelnes, J.E. and Knudsen, L.E., Stainless steel welding and semen quality, *Reprod. Toxicol.*, 2, 213, 1988.
156. Rachootin, P. and Olsen, J., The risk of infertility and delayed conception associated with exposures in the Danish workplace, *J. Occup. Med.*, 25 (5), 394, 1983.
157. Kenkel, S., Rolf, C., and Nieschlag, E., Occupational risks for male fertility: an analysis of patients attending a tertiary referral centre, *Int. J. Androl.*, 24 (6), 318, 2001.

chapter 9

Use of Metal Reproductive Toxicity Data in Selecting Ecological Toxicity Reference Values for Small Mammals Inhabiting Hazardous Waste Sites

Michael J. Anderson, Julie T. Yamamoto, and Hilary Waites
Office of Spill Prevention and Response,
California Department of Fish and Game

Contents

Introduction

Ecological risk assessment (ERA) has gained broad regulatory application as a standardized yet flexible approach to managing both intentional and unintentional environmental chemical releases. ERA is scalable to both small, relatively simple scenarios, and larger, complex environmental problems. As with human health risk assessment, ERA is essentially based on the scientific method whereby hypotheses (about toxicity or other hazards) are developed and tested, in iterative fashion, until a risk conclusion can be developed with an acceptable level of certainty. This risk characterization can then be used with other appropriate factors in decision making during site remediation and other management activities. Unlike human health risk assessment, ecological risk assessment must be capable of addressing multiple receptor types, including plants, invertebrates, and vertebrates. Exposure assumptions must be tailored to the specific life histories and biology of each receptor in order for the risk assessment to be valid. Similarly, to the extent possible, toxicity reference values used in the assessment must be reasonably expected to protect each representative receptor of concern. For example, benchmarks for plants must be developed from plant toxicity data, whereas those for mammals must be derived from mammalian toxicity data. Depending on the chemical and receptor species of concern, the availability of toxicological data may be sparse, which often leads to a significant source of uncertainty in the risk characterization.

Guidelines published by the U.S. Environmental Protection Agency (USEPA) form the basis for many types of contaminated site cleanup actions (USEPA 1997, 1998). The process for ERA can be divided into three main phases:

1. Problem Formulation/Conceptual Model Development: Using basic site information, potential chemicals and receptors of concern are identified and conceptual models of potential exposure pathways, including food-chain transport, are developed. Assessment goals and endpoints are developed.
2. Analysis
   a. Exposure: Develop qualitative and quantitative exposure estimates for receptors of concern; information collected includes concentrations of chemicals of concern in exposure media such as air, water, soil, sediment, and tissues.
   b. Effects: Develop appropriate toxicity reference values for contaminants of concern, considering receptors of concern and ecologically relevant toxicity endpoints. This analysis can include literature searches, *in situ* toxicity testing, laboratory bioassays, or population- and community-level assessments.
3. Risk Characterization: Integrate exposure and effects analyses to characterize risks to receptors of concern. This phase also considers uncertainties in the risk assessment, including information gaps, potential errors in conceptual modeling, and extrapolation uncertainty.

Toxicity Reference Values

Assessing ecological risk to a species inhabiting a hazardous waste site requires the development of an appropriate toxicity reference value for each chemical pollutant of concern (Department of Toxic Substances Control [DTSC] 1996a, DTSC 2000, USEPA 1997). A toxicity reference value (TRV) is the daily dose of a chemical (e.g., mg chemical/kg wet body weight/day) that elicits a particular biological effect. When evaluating potential ecological risks, toxicity endpoints commonly considered to be ecologically relevant are mortality, growth, reproduction, and development. Nevertheless, subtle behavioral, histopathologic, neurologic, immunologic, and organ-specific effects should also be considered in TRV development, inasmuch as these endpoints could potentially affect the survival or reproductive fitness of an animal. Application of TRVs in an ERA typically takes the form of a hazard quotient (HQ), whereby the ratio of the estimated daily exposure of the species of concern to the TRV daily dose is used to evaluate likelihood of toxicity. Clearly, the risk evaluation can be largely influenced by the selected TRV.

Example of TRV Derivation: Lead

Lead (Pb) is frequently cited as a contaminant of potential ecological concern in soil at hazardous waste sites, especially at Department of Defense facilities. In order to evaluate ecological risks to small mammals living in Pb-contaminated areas, we have developed a protective TRV for use in screening-level ERAs (e.g., Tier I, USEPA 1997). A relatively large literature database exists regarding the toxicity of Pb in laboratory rodent species. Shore and Douben (1994) concluded that these species are acceptable surrogates for predicting Pb hazards to wild mammals, but that direct extrapolation of laboratory dose-response data to the wild situation could not be done. This conclusion is based on the fact that the bioavailability of lead in the field is often markedly different than in the laboratory, and the toxicological sensitivity of wild animals may encompass a greater range than demonstrated in laboratory species. Nevertheless, regulatory agencies and responsible parties must have protective and scientifically defensible TRVs for making a decision to proceed to a baseline risk assessment (e.g., Tier II, USEPA 1997; Phase II Validation Study, DTSC 1996a) and to develop ecologically protective cleanup goals.

Lead Literature Review

We surveyed the available secondary literature (e.g., Agency for Toxic Substances and Disease Registry [ATSDR] 1999) and primary Pb literature sources to identify the lowest ecologically relevant No Observable Adverse Effects Levels (NOAELs) and Lowest Observable Adverse Effect Levels (LOAELs) from oral exposure to Pb. This approach is supported by the Superfund guidance for ecological risk assessment (USEPA 1997), meets the objectives of a Tier I or screening-level risk assessment, and is protective of individuals that are potentially the most toxicologically sensitive to Pb. While no single study can account for differences in species susceptibility and differences in effects and absorption for various forms of Pb, a single study can be used to represent a dose that is protective for a variety of species and toxicological endpoints (i.e., protecting the most toxicologically sensitive small mammal species is protective of other small mammal species).

We focused our literature review on studies on small mammals (mostly laboratory rats and mice) where Pb, in various chemical forms, was administered orally over an intermediate (greater than 14 days) or chronic (greater than 365 days) exposure period (ATSDR 1999). Acute exposure studies (other than exposures to Pb during gestation), and studies utilizing other exposure routes, including injection, dermal absorption, and inhalation, were not considered in TRV development. Small mammal studies, as opposed to large mammal studies (i.e., dogs, cats, lambs, cattle, pigs, monkeys, humans), were preferred because these species are typically the most exposed receptor populations at hazardous waste sites due to their burrowing behavior and small foraging ranges. Many studies were rejected from consideration, including those studies where (1) conclusions were drawn using inappropriate statistics, (2) the endpoints evaluated had questionable ecological relevance or presented difficulty in interpretation (i.e., aminolevulinic acid dehydratase [ALAD] changes, hepatic glutathione levels, serum chemistry data, blood pressure changes), (3) a clear dose–response relationship was not demonstrated, and/or (4) it was not possible to reconstruct or determine the dose or form of Pb administered.

One hundred nine studies were selected from the ATSDR *Toxicological Profile for Lead* (ATSDR 1999) that were based on the oral administration of Pb to small mammals (*i.e.*, laboratory rats and mice). Among the NOAEL and LOAEL values reported in the ATSDR profile, it was apparent that the lowest NOAELs and LOAELs ranged primarily between 0.1 and 10 mg/kg BW/day. Studies with LOAELs greater than 10 mg/kg/d were eliminated from further consideration (47 studies). In addition, many of these studies included endpoints that were of questionable ecological relevance or presented difficulty in interpretation, such as blood pressure, enzyme, receptor, and neurotransmitter concentrations in circulation or tissues (18 studies). Other studies did not contain adequate information on dose and route. Using these considerations, 30 studies were selected for further review in the TRV selection process (i.e., NOAEL ≤ 1; LOAEL ≤ 10 mg/kg BW/day) (Table 9.1).

Each study was evaluated for the following:

- experimental design and toxicity endpoint(s)
- exposure duration/frequency
- Pb chemical form
- exposure medium
- test species and strain
- observed effects (including statistical inferences)
- best professional judgment concerning the value of the paper, with focus on fatal flaws, if any were identified

Table 9.1 Summary and Critical Review of Lead Toxicity Studies in which NOAELs Are Reported as Less Than or Equal to 1 mg/kg BW/day.

Test Species[1]	Elemental Pb Conc. in Food or Water[2]	Chemical Form	LOAEL[3] (mg/kg BW/day)	NOAEL[3] (mg/kg BW/day)	Toxicity Endpoint	Notes/Professional Judgment	Reference
Rat (42- to 49-day-old males, F344/N)	0, 10, 30, and 100 mg/kg	Pb Ac	5[a]	0.5[a]	Body weight loss, urinary aminolevulinic acid excretion, kidney histopathological lesions	Effects of Pb acetate on body weight loss and kidney consistent with findings of Fowler et al. (1980) and Khalil-Manesh et al. (1993). Calculation of dose is subject to uncertainty because body weights of tested rats are not reported. Study shows that chemical form of Pb (i.e., Pb Ac, Pb S, Pb Ore) clearly affects bioavailability. This study supports a 1 mg Pb/kg BW/day NOAEL TRV for kidney effects.	Dieter et al. 1993
Rat (100-day-old females, Wistar)	0, 9.3 mg/kg[b]	Pb Ac	0.9[a]	ND	Bone structure, density, length	Only one dose was tested and the relationship between the toxicity endpoint and an adverse ecological effect is not strong. The study should not be used to set a TRV; however, it does support other studies that show more relevant effects at similar dose levels.	Escribano et al. 1997
Rat (Adult)	0, 55, and 2750 mg/L[b]	Pb Ac	6[a]	ND	Bone density and histology	Decrease in bone density noted in the lowest treatment group (55 ppm Pb) after 12 months exposure. Relationship between toxicity endpoint and adverse ecological effect not strong. The study should not be used to set a TRV; however, it does support other studies that show more relevant effects at similar dose levels.	Gruber et al. 1997

Table 9.1 Summary and Critical Review of Lead Toxicity Studies in which NOAELs Are Reported as Less Than or Equal to 1 mg/kg BW/day.

Test Species[1]	Elemental Pb Conc. in Food or Water[2]	Chemical Form	LOAEL[3] (mg/kg BW/day)	NOAEL[3] (mg/kg BW/day)	Toxicity Endpoint	Notes/Professional Judgment	Reference
Rat (female, Sprague-Dawley)	0, 0.1, 1, and 10 mg/L[c]	Pb N	1[c,d]	0.1[c,d]	Decrease in ALAD activity in blood and kidney. Increase in free tissue porphyrins in kidney of newborns (21 day old)	Only four rats per dose level. No effects on fertility, gestation, viability, lactation, or blood biochemistry observed in adult females. No behavioral, morphologic, or histologic endpoints evaluated, so there are limited inferences that can be made about functional effects. The study should not be used to set a TRV; however, it does support other studies that show kidney effects at low dose levels.	Hubermont et al. 1976
Rat (35 days old)	N/A	Pb Ac	0.4[e]	ND	Neurological: changes in G Protein signal transduction	Only prenatal effects observed; no adult effects. See below for Singh and Ashraf, 1989.	Singh 1993
Rat (35 days old)	N/A	Pb Ac	0.4[k]	ND	Neurological: changes in amino acid and neurotransmitter concentrations	Methods not rigorously presented. Effects more pronounced in developing young; however, difficult to equate changes in neurotransmittor levels with ecological relevance. The study should not be used to set a TRV; however, it does support other studies that show behavioral effects at low dose levels.	Singh and Ashraf 1989
Rat (35 to 42 days old, male Buffalo)	0, 19 mg/kg BW once a week, 39 mg/kg BW twice a week[b]	Pb Ac	5	ND	Atrophy of the elastic fibers of the aorta, decrease in serum cholesterol, and 24% increase in serum triglycerides	Study was designed to investigate the effects of Pb on the development of atherosclerosis in humans. No dose produced clinical signs of Pb poisoning. The study should not be used to set a TRV because there is questionable ecological relevance of the endpoints considered.	Skoczynska et al. 1993

Species	Dose	Chemical	Value		Endpoint	Notes	Reference
Rat (21 to 56 days postpartum males, Sprague-Dawley)	0, 11.4, 99.7 mg/L[f]	Not reported[f]	1.3[a]	ND	Increase in tissue lead retention; significant reduction in choice behavior in a complex maze	Sensitive life stage with ecologically relevant endpoints. Single generation study with no maternal exposure; really a study of zinc deficiency. It was concluded that because the maze choice behavior effect was not linear, only 3 doses were tested, and because a NOAEL dose was not determined, the study should only be considered as supporting evidence for establishing a NOAEL TRV near 1 mg Pb/kg BW/day.	Bushnell and Levin 1983
Rats (weanling male and female, Long-Evans)	0, 28, 83 mg/L[b]	Pb Ac	3[a]	ND	Altered cholinergic sensitivity in response to lead	Study was designed to investigate the effects of various muscarinic agonists and antagonists on rats chronically exposed to Pb. The study should not be used to set a TRV because there is questionable ecological relevance of the endpoints considered.	Cory-Slechta and Pokora 1995
Rats (weanling male and female, Long-Evans)	0, 28, 55, 275 mg/L[b]	Pb Ac	3[a]	ND	Increases in the rate of fixed-interval lever-response and running time	Marked individual differences in susceptibility to Pb-induced increases in performance on a fixed interval 1-min schedule of food reinforcement. The study should not be used to set a TRV because there is questionable ecological relevance of the endpoints considered.	Cory-Slechta et al. 1985
Rats (weanling male and female, Long-Evans)	0, 14 mg/L[b]	Pb Ac	1.5[a]	ND	Increases in the rate of fixed-interval lever-response and running time	See Cory-Slechta et al. 1985 above.	Cory-Slechta et al. 1983

Table 9.1 Summary and Critical Review of Lead Toxicity Studies in which NOAELs Are Reported as Less Than or Equal to 1 mg/kg BW/day.

Test Species[1]	Elemental Pb Conc. in Food or Water[2]	Chemical Form	LOAEL[3] (mg/kg BW/day)	NOAEL[3] (mg/kg BW/day)	Toxicity Endpoint	Notes/Professional Judgment	Reference
Rats (70 days old, male, Wistar)	0, 0.2, 18, 182 mg/kg[b] BW	Pb Ac	0.2	ND	Reduction in the total number of spermatozoa in testes Decrease in body weight gain and % normal spermatozoa	The decrease in sperm numbers was not dose related and only statistically different from the control at the lowest dose tested. Decrease in body weight and % normal spermatozoa only found at the highest dose tested. The study should not be used to set a TRV because of the lack of a dose response relationship for the most sensitive endpoint examined (i.e., number of spermatozoa).	Barratt et al. 1989
Rat (F1, 40 days old, Sprague-Dawley)	0, 25, 50 mg/L	Pb Ac	5.6	ND	Immunotoxicity, including negative effects on cell-mediated immune function and reduced thymus weight	Dose estimated from companion study Kimmel et al. 1980. The Fowler et al. (1980) study should be used to set the TRV in absence of NOAEL for this study. Study supports a 1 mg Pb/kg BW/day NOAEL TRV.	Faith et al. 1979
Rat (F1, 270 days old, males, Sprague-Dawley)	0, 0.5, 5, 50, 250 mg/L	Pb Ac	5.6	1	Adverse histologic lesions in male kidney (cyto- and karyomegaly)	Doses estimated from companion study Kimmel et al. 1980. Viral infection in one replicate; however, data are excluded from analysis. This study (and associated companion studies, including Luster et al. 1978, Faith et al. 1979, Grant et al. 1980, and Kimmel et al. 1980) support a NOAEL TRV of 1 mg/kg BW/day for kidney related adverse effects, which is protective of other reproductive, immunologic, and behavioral effects seen a similar or higher dose levels.	Fowler et al. 1980

Species/strain	Concentration	Compound	Value	UF	Endpoint	Comments	Reference
Rat (F1, 30 to 60 days old, Sprague-Dawley)	0, 0.5, 5, 50, 250 mg/L	Pb Ac	5.6	1	Physical and behavioral development, delayed surface and air righting	Doses estimated from companion study Kimmel et al. 1980. Study supports a 1 mg Pb/kg BW/day NOAEL for behavioral adverse effects.	Grant et al. 1980
Rat (F0, F1 fetuses, Wistar)	N/A	Pb Ac	0.5g	ND	Decreased erythrocyte ALAD activity in pups; lower fetal weights	Changes in ALAD activity not ecologically relevant. Reduced body weight of fetuses or pups not reported in other studies utilizing a similar low dose of Pb Ac. This study should not be used to generate a TRV.	Hayashi 1983
Rat (F1, 30 to 60 days old, females, Sprague-Dawley)	0, 0.5, 5, 25, 50, 250 mg/L	Pb Ac	5.6	1	Maternal toxicity and perinatal growth, delayed vaginal opening	High variability of vaginal opening data and inconclusive link to adverse effect (see study). The Fowler et al. 1980 companion study should be used to set the TRV.	Kimmel et al. 1980
Rat (F1, 40 days old, Sprague-Dawley)	0, 25, 50 mg/L	Pb Ac	5.6	ND	Immunotoxicity, including negative effects on antigen stimulation, antibody production, and thymus weight	Dose estimated from companion study Kimmel et al. 1980. The Fowler et al. 1980 study should be used to set TRV in absence of NOAEL for this study.	Luster et al. 1978
Rat (2 generations, Sprague-Dawley)	0, 5, 50 mg/L	Pb Ac	0.7	ND	Delayed eye opening and righting reflex development, and transient decreased exploratory activity in treatment groups	The ecological relevance of the endpoints measured is questionable. Similarly, the biological significance of the minimal delays and transient hypoactivity is also questionable. This study should not be recommended as the basis for setting the Pb TRV.	Reiter et al. 1975

Table 9.1 Summary and Critical Review of Lead Toxicity Studies in which NOAELs Are Reported as Less Than or Equal to 1 mg/kg BW/day.

Test Species[1]	Elemental Pb Conc. in Food or Water[2]	Chemical Form	LOAEL[3] (mg/kg BW/day)	NOAEL[3] (mg/kg BW/day)	Toxicity Endpoint	Notes/Professional Judgment	Reference
Rat (F1, 150 days old, males, Charles River)	0, 5, 25 mg/L	Pb Ac	2.2	ND	Basal decreases in plasma renin concentration (highest dose) and in renal renin concentrations (low and high doses) No effects on systolic blood pressure	Author states that the most biologically significant effects occurred in the 25 ppm exposure group (LOAEL shown). This study should not be recommended as the basis for setting the Pb TRV because the linkage between altered sensitivity of the renin-angiotension system and ecologically relevant effects is tenuous. Functional tests of or histopathological lesions in the kidney were judged to be more significant, ecologically relevant effects (i.e., Fowler et al. 1980).	Victery et al. 1982
Rats (male and female)	0, 10, 50, 100, 500, 1000, 2000 mg/kg	Pb Ac	10 mg/kg increased stippled cell; 50 mg/kg decreased ALAD; 500 mg/kg increased kidney tumors in male; 1000 mg/kg reduced weight gain; 1000 mg/kg reduced weanling weight	10 mg/kg no effect on ALAD; 500 mg/kg no effect on weanling weight	Growth and mortality; blood parameters, ALAD; microscopic examination for kidney tumors; Number of pregnancies, pups; fertility, gestation, viability, and lactation indices; weanling weight and kidney histology	Adverse effects associated with reduced ALAD are unknown, as well as ecological relevance. The strength of this study is that it involves a three-generation, six-litter reproductive study with seven different dose levels. Reduced weanling weight has important implications for reproductive potential in the wild; however, effects on body weight were only apparent at doses estimated to be greater than 10 mg Pb/kg BW/day.	Azar et al. 1973

Species	Dose	Compound			Endpoint	Comments	Reference
Guinea pig (females at days 22–62 of pregancy)	0, 5.5, and 11 mg/kg/d	Pb Ac	5.5	ND	Reduction in hypothalamic levels of GnRH and SRIF	The ecological relevance of the endpoints measured is questionable. No function tests were performed to assess significance of changes in peptide hormone levels. This study should not be used to set the Pb TRV-low; however, it is supportive of adverse reproductive effects occurring at similar or higher doses.	Sierra and Tiffany-Castiglioni 1992
Mouse (male weanlings, Balb-C albino Swiss)	Unknown	Pb Monoxide Alloy	0.1 mg/animal/day	ND	Decreased litter size and numbers of spermatogenic cells in seminiferous tubules	Neither body weights nor alloy composition were provided. The Pb alloy caused decreased reproductive performance in male mice with some evidence of a dominant lethal effect at the highest dose. However, the lack of analysis of the test article leaves uncertain which other metals might have been present in the diets of the test animals. The lower dose also caused an adverse effect on implantation and litter size. The lack of chemical analysis of the alloy precludes assigning this toxicity to Pb. This study cannot be used to set a TRV for Pb.	Al-Hakkak et al. 1988
Rat (female, through pregnancy, Sprague-Dawley)	0, 10, 50, 100, 200, 500 mg/L	Pb Ac	7[a]	1.3[a]	Maternal weight gain, feeding efficiency, litter size, and fetal weight	Short-term study undertaken during a critical life stage (days 1 to 21 of gestation) and may be considered short-term chronic, at least for these endpoints. The NOAEL for reproduction supports the proposed 1 mg Pb/kg BW/day NOAEL TRV.	Dilts and Ahokas 1979

Table 9.1 Summary and Critical Review of Lead Toxicity Studies in which NOAELs Are Reported as Less Than or Equal to 1 mg/kg BW/day.

Test Species[1]	Elemental Pb Conc. in Food or Water[2]	Chemical Form	LOAEL[3] (mg/kg BW/day)	NOAEL[3] (mg/kg BW/day)	Toxicity Endpoint	Notes/Professional Judgment	Reference
Rat (female, F0–150 days old F1, Sprague-Dawley)	0, 20, 200 mg/L	Pb Cl	2.6[a]	ND	Reduction in uterine estradiol receptors and increases in receptor affinity	*Ex vivo* estradiol-binding experiments cannot be readily used to quantify the likelihood of adverse effects on ecological receptors. However, decreases in receptor number could eventually affect follicular responsiveness to estrogen, potentially resulting in an adverse reproductive effect. This study should not be used to set a TRV; however, it is supportive of potential adverse reproductive effects occuring at these and higher dose levels.	Wiebe and Barr 1988
Mouse (female adult, albino Swiss)	0, 2, 4, and 8 mg/kg/d	Pb Ac	2	ND	Number of small, medium, and large follicles reduced in ovary	Pb was detected in the control group (21.7 ug/dl blood, see Table 3 in the study), which limits inferences that can be made from the study. Exposure period was 60 days, but gavages were administered only 5 days per week. Large ovarian follicles showed a statistical effect (ANOVA) at 4 mg/kg BW/day, but small and medium ovarian follicles showed a statistical effect at 2 mg/kg BW/day. Given the short, and irregular, exposure period it is likely that the LOAEL would be lower if dosing were continued for a longer period on a daily basis. This study is supportive of adverse reproductive effects caused by low level exposure to Pb, but inappropriate to set a TRV.	Junaid et al. 1997

Mouse (adult males and females, through pregnancy, LACA strain)	0, 6.0, 14, and 28 mg/kg[b]	Pb Ac	6	ND	Living embryos per mother significantly reduced, kidney activity of glutathione increased and alkaline phosphatase significantly decreased from control levels in pregnant mice, kidney accumulation of lead greater in nonpregnant females	The small number of animals used in this study limits its usefulness as a primary source for setting a Pb TRV. Renal effects reported were not related to functional deficits in the kidney. However, based on the reproductive effects observed, the study supports a 1 mg Pb/kg BW/day NOAEL TRV.	Gupta et al. 1995
Cotton Rat (adult males and females, wild)	0, 55, 550 mg/L[b]	Pb Ac	11[a]	ND	Compromised cellular immune and hematological responses, kidney lesions, reduction in liver, seminal vesicle mass, testes sperm content, and ovarian follicles	Histopathologic lesions consistent with Pb toxicosis, including altered renal proximal tubular epithelium, renal intranuclear inclusions, and at the highest dose, lowered numbers of sperm and developing follicles. This study is supportive of a 1 mg Pb/kg BW/day NOAEL TRV.	McMurray et al. 1995
Rat (60 days old, male, Sprague-Dawley)	0, 55 mg/L[b]	Pb Ac	6[a]	ND	Transient increase in glomerular filtration rate, increase in urinary n-acetyl-[beta]-D-glucosaminidase, tubular atrophy, and interstitial fibrosis (at 365 days)	Authors state that exposure produced no significant changes in renal function; however, mild histopathologic lesions were evident at 12 months, suggesting incipient damage to the proximal tubules. Study was not as long in duration as Fowler et al. (1980), did not include a sensitive life stage (e.g., fetal, weanling), and only two doses were evaluated. This study supports a 1 mg Pb/kg BW/day NOAEL TRV.	Khalil-Manesh et al. 1993

Table 9.1 Summary and Critical Review of Lead Toxicity Studies in which NOAELs Are Reported as Less Than or Equal to 1 mg/kg BW/day.

Test Species[1]	Elemental Pb Conc. in Food or Water[2]	Chemical Form	LOAEL[3] (mg/kg BW/day)	NOAEL[3] (mg/kg BW/day)	Toxicity Endpoint	Notes/Professional Judgment	Reference
Rat (55–60 days old, pregnant, Long-Evans; pups exposed via breast milk until weaning)	0, 109, 1090 mg/L[h]	Pb Ac	12.6[a]	ND	Most significant effects found in pups, including retinal degeneration; loss of rod cells; reduction in rod cell sensitivity; reduced rhodopsin content; reduced cGMP phosphodiesterase activity; and reduced cGMP concentration in retina	Loss of visual acuity, particularly night vision, seen as a significant adverse ecologically relevant effect. Lowest effects dose estimated in *dames* that causes adverse effects in pups is slightly greater than 10 mg/kg BW/day. This study supports a 1 mg Pb/kg BW/day NOAEL TRV.	Fox et al. 1997

FOOTNOTES

Pb Ac = lead acetate
Pb S = lead sulfate
Pb N = lead nitrate
Pb O = lead oxide

BW = body weight
GD = gestational day
MOA = Mode of administration
LOAEL = lowest observable adverse effect level
NOAEL = no observable adverse effect level
N/A = not applicable or not available
N/D = not determined

1. Parenthetical comments include age of test species when significant effects were observed, gender, and strain of test species evaluated.
2. Elementation concentrations of lead in food or water tested in the study, unless otherwise noted.
3. Dose reported in the study or estimated using average body weight and intake rates provided in the study.

a. Dose estimated because BWs, drinking water rates, and/or feed intake rates are not reported in the study.
b. Elemental concentration of lead determined assuming atomic weight of Pb is 207.2 and molecular weight of Lead(II) Acetate Trihydrate (commercially available form of lead) is 379.33. Therefore, elemental concentration of lead administered assumed as 55% of the total mass of lead acetate trihydrate.
c. Elemental concentration reported by the authors; however it is not clear in the methods whether concentration reported reflects elemental Pb or Pb nitrate. Therefore, elemental concentration of lead administered assumed as 63% of the total mass of lead nitrate.
d. Maternal dose through pregnancy resulting in adverse effects on pup.
e. Dose reported in the study, converting lead acetate dose to elemental lead dose as described in footnote c above and assuming that doses were administered 5 days per week.
f. Form of lead administered not reported, however actual concentrations of elemental lead (by atomic absorption assay) in drinking water were reported.
g. Dose reported by ATSDR (1999).
h. Elemental Pb concentration reported in the study.

The purpose of the review was to determine the best study, or best set of studies, to support and represent a small mammal *no effect* and *effects-based* TRV for Pb, as discussed in more detail below.

Summary of the Studies Selected for TRV Development

A series of papers was published that examined multiple effects on rats exposed to Pb during the same experiment. Effects on postnatal physical and behavioral development (Grant et al. 1980), maternal toxicity and peri-natal effects (Kimmel et al. 1980), immunological effects (Luster et al. 1978; Faith et al. 1979), and renal effects (Fowler et al. 1980) were described. Of these studies, that of Fowler et al. (1980) was deemed to be the most appropriate for establishment of a NOAEL- and LOAEL-based TRV for Pb.

Fowler et al. (1980) reported the adverse effects of Pb on renal structure and mitochondrial function. Male and female Sprague-Dawley rats were exposed to 0, 0.5, 5, 25, 50, and 250 mg Pb (as Pb acetate)/L in drinking water during gestation, lactation, and the first 6 or 9 months of life. Tissue Pb concentrations and ALAD were increased in a dose-dependent manner. Renal tubule mitochondrial respiration was depressed in a dose-dependent manner by Pb treatment. Kidney weight and the incidence of histopathologic lesions in the proximal tubule were slightly increased (in males only) relative to controls in the 5 mg Pb/L treatment group. Histopathologic lesions included enlarged tubular lining cells (cytomegaly) with enlarged nuclei (karyomegaly). The decreases in mitochondrial respiration supported the observed cytomegaly and karyomegaly, as well as the fact that others have associated renal proximal tubular dysfunction with impairment of mitochondrial function (see Goyer et al. 1972). No statistical testing was conducted to evaluate the semiquantitative histopathological scoring data; however, the data do suggest a dose-related increase in the severity of proximal tubule cytomegaly, inclusion bodies (primarily Pb precipitates), and hemosiderin content.

Fowler et al. (1980) concluded that the 5 mg Pb/L treatment group was the lowest Pb drinking water concentration that elicited a detectable effect (median blood concentration of 11 μg/dL). Coauthors Grant et al. (1980) considered the renal changes at 5 mg Pb/L as "subtle." Furthermore, Grant et al. (1980), Kimmel et al. (1980), and Fowler et al. (1980) considered 25 mg Pb/L as the lowest concentration that produced organ-specific toxicity (median blood concentration of 21 μg/dL). Given the uncertainty associated with changes in enzyme levels (i.e., mitochodrial respiration) as being indicative of an adverse effect, the minimal cytomegaly reported in male kidneys after 9 months, and lack of organ-specific or behavioral toxicity associated with the 5 mg Pb/L treatment group, we concluded that the 5 mg Pb/L drinking water concentration best represented a NOAEL and the 25 mg Pb/L concentration best represented a LOAEL. Using female dosing information provided in Kimmel et al. (1980, Table 1 of the publication), we estimated a 95% upper confidence limit of the mean NOAEL (1 mg Pb/kg BW/day) as protective of potentially adverse renal effects, and a LOAEL of 5.6 mg Pb/ kg BW/day as indicative of potentially adverse renal effects. Based on the strength of the study design, we concluded that the findings of the Fowler et al. (1980) and associated studies should form the basis of a NOAEL- and LOAEL-based TRV for Pb.

Summary of Studies Considered Supportive of a TRV
We identified 21 studies that directly or indirectly support the development of a mammalian Pb TRV. Doses near or below 10 mg/kg BW/day cause a variety of potentially adverse effects, including changes in bone density or structure (i.e., osteodystrophy), behavior, immune function, reproductive capacity, embryonic development, and renal ultrastructure. A maternal dose of 12.6 mg Pb/kg BW/day causes multiple biochemical and functional changes in the eye (i.e., ocular changes) of lactationally exposed pups, including degeneration of the retina. Each effect is discussed in more detail below.

Reproductive/Developmental Effects.   Seven studies, Dilts and Ahokas (1979), Kimmel et al. (1980), Wiebe and Barr (1988), Sierra and Tiffany-Castiglioni (1992), Gupta et al. (1995), McMurry et al. (1995), and Junaid et al. (1997), suggest either directly or indirectly that low doses of Pb affect reproduction and sexual development in small mammals.

Dilts and Ahokas (1979) exposed pregnant female Sprague-Dawley rats to 0, 10, 50, 100, 200, or 500 mg Pb (as Pb acetate)/L in drinking water over a 21-day gestation period. There were dose-dependent decreases in food ingestion rate (50 mg Pb/L) and both total (50 mg Pb/L) and net (total minus conceptus) body weight (100 mg Pb/L), but drinking rate was not significantly different from control. There was no difference in live litter size, or number of litters with dead fetuses, but the number of dead fetuses increased with concentration. The live litter weight (50 mg Pb/L) and average fetal weight (100 mg Pb/L) deceased with increasing concentration. The adverse effects on food ingestion and net maternal weight at 50 mg Pb/L introduces a high degree uncertainty of the cause of fetal effects, because fetal nutritional status may have varied directly with maternal nutritional status. However, the lack of effects on placental weight suggests that Pb toxicity was the major determinant. Body weights were not reported by the authors, but time-dependent maternal weight was presented in Figure 1 of the study. The maternal body weight at the start of the experiment for the 50 mg Pb/L treatment group was approximately 245 g. Using the water ingestion rates provided by the authors (34.2 mL/day), the LOAEL for reproductive effects is 7 mg Pb/kg BW/day. Using the water ingestion rate for the next lower dose (37 mL/day) yields a NOAEL for reproductive effects of 1.3 mg Pb/kg BW/day. Although these were short-term studies, they were undertaken during a critical life stage (days 1 to 21 of gestation) and may be considered short-term chronic, at least for these endpoints. The NOAEL of 1.3 mg Pb/ kg BW/day supports the selected TRV.

Kimmel et al. (1980) exposed Sprague-Dawley rats to 0, 0.5, 5, 25, 50, or 250 mg Pb (as Pb acetate)/L for 6 to 7 weeks premating and continuously through gestation and lactation. No changes in food or water consumption were noted. No effects were found on the ability to conceive, carry normal litter to term, or deliver the young. No effects were found on the percentage of malformed fetuses, resorptions, or postpartum pup deaths; however, body lengths were significantly reduced in the 250 mg Pb/L treatment group. The reduction in body length without reductions in food or water consumption could not be explained, but may relate to findings of Escribano et al. (1997) and Gruber et al. (1997) described below. Vaginal opening was significantly delayed in the 25 mg Pb/L treatment group; however, there was a high degree of variability in the data and no link to an adverse biological effect (i.e., delay did not affect the rats' ability to reproduce). Furthermore, in a more recent two-generation Sprague-Dawley rat reproductive study, Ronis et al. (1998) showed that Pb-related differences in vaginal opening and prostate weight disappeared after 85 days. Ronis et al. determined that the likely cause of the effects was that Pb suppressed the normal sex steroid surges observed at birth and during puberty. Using body weight and intake rate information provided by the authors of the Ronis et al. (1998) study, we estimated a NOAEL of 32 mg Pb/kg BW/day for reproductive effects. From the Kimmel et al. (1980) study, we estimated a NOAEL of 1 mg Pb/kg BW/day for similar reproductive effects (e.g., vaginal opening, see Table 9.1. Both authors stated that the pH of the administered drinking water solution was adjusted such that Pb was not observed to precipitate from solution. Hence, the reason for the disparity in drinking water effect–levels for vaginal opening in the Kimmel et al. (1980) and Ronis et al. (1998) studies is unknown, but it may relate to differences in Pb absorption, sequestration, or elimination among test animals used in each study. This hypothesis is supported by the fact that in both studies the lowest effective blood Pb level that induced sexual maturation effects in juvenile rats was approximately 20 µg/dL. The ecological significance of delayed vaginal opening or sexual maturation, if it occurred in wild rats, is unknown.

Wiebe and Barr (1988) exposed Sprague-Dawley rats to three different Pb exposure regimes. Female rats were exposed to 0, 20, or 200 mg Pb (Pb as Pb chloride)/L in drinking water for 118 days prior to mating, and the offspring were sacrificed after 21 days (Experiment 1). There were no effects on the number of uterine estradiol (E2) receptors of 21-day-old offspring, but the affinity of the receptors was reduced. Female rats were exposed to 0, 20, or 100 mg Pb/L beginning on day 7 of pregnancy and continuing until the offspring were weaned (Experiment 2). Half of the offspring from each group were exposed to 20 mg Pb/L and half received no additional Pb exposure. Exposures continued until the rats were 150 days old. A significant decrease in the number of uterine E2 receptors was observed. Twenty-one-day-old rats were exposed to 0, 20, or 200 mg Pb/L for 35 days (Experiment 3). There were no effects on uterine weight or, except for a brief period, whole body weight. Pb was associated with increases in receptor affinity for E2 (Experiments 1, 2, and 3) and with either no change (Experiment 1) or a decrease in the number of E2 receptors (Experiments 2 and 3), regardless of age at exposure. However, variance was high and there was no dose response beyond that found in the 20 mg Pb/L treatment group. The authors stated that changes in uterine E2 receptor numbers and affinity may occur maximally at low, subclinical exposures to Pb. We did not consider that *ex vivo* estradiol-binding experiments could be readily used to evaluate the likelihood of adverse effects on wildlife. However, decreases in receptor number could eventually affect follicular responsiveness to estrogen, potentially resulting in an adverse reproductive effect. A LOAEL of 2.6 mg Pb/ kg BW/day day was estimated for the study. We concluded that this study should not be used to set a TRV; however, it was supportive of potential adverse reproductive effects resulting from low developmental Pb exposure.

Sierra and Tiffany-Castiglioni (1992) gave pregnant guinea pigs a daily oral dose of 0, 5.5, and 11 mg Pb (as Pb acetate) per kg BW. Hypothalamic levels of gonadotropin releasing hormone and somatostatin were reduced in a dose-dependent manner in both dams and fetuses. However, neither litter size nor body and organ weights, including placental weights, of the dams and fetuses were affected. The authors postulated that a reduction in peptide hormone levels may alter *in utero* sexual development, resulting in developmental effects that do not become apparent until sexual maturity or breeding (i.e., F1 generation endpoints not evaluated in the study). We concluded that this study should not be used to set a TRV; however, it was supportive of potential adverse effects resulting from low-level maternal and developmental Pb exposure.

Gupta et al. (1995) exposed two-month-old female mice (LACA strain) to daily doses of 0, 6.0, 14, and 28 mg Pb per kg BW by gavage through pregnancy. Females were treated with human chorionic gonadotropin (hCG) and allowed to mate with non-Pb treated males. The number of living embryos conceived by the Pb-treated females was significantly reduced (29%) from controls, but a weak dose–response relationship was demonstrated. The poor dose–response relationship may have been the result of the low numbers of animals in each treatment group (i.e., less than six animals). We considered this study as supportive evidence for the adverse reproductive effects of Pb at low doses, but inappropriate to set a TRV.

Junaid et al. (1997) exposed female Swiss albino mice to daily doses of 0, 2, 4, or 8 mg Pb (as Pb acetate)/kg BW/day for 60 days (5 days/week, gavage). A clear dose–response relationship was established. The lowest dose significantly reduced the number of small, medium, and large follicles present in ovaries. Rates of follicular atresia were also increased, but the effect was only observed at the highest dose for medium-sized follicles. An effects-level of 2 mg Pb/kg BW/day was reported by the authors. However, it was not clear whether the effects-level was expressed as elemental Pb and adjusted for the 5 days per week exposure protocol. The study is confounded by the fact that the control animals received significant Pb exposure (i.e., greater than 20 µg Pb/dL blood) from an unknown source (e.g., contaminated feed or drinking water). We considered this study as supportive evidence for the adverse reproductive effects of Pb, but inappropriate to set a TRV.

McMurry et al. (1995) exposed wild adult male cotton rats to 0, 55, or 550 mg Pb/L in drinking water for 7 or 13 weeks. When rats were exposed over a 7-week period that included maturation of the gonads (i.e., breeding season), Pb at the highest dose reduced seminal vesicle and epididymis mass. Testicular and ovarian histological lesions were detected in animals exposed to 55 mg Pb/L or 550 mg Pb/L; lesions were usually more frequent and severe at the higher dose. Lesions included reductions in spermatogenesis in males and the lack of developing follicles in females. We estimated a LOAEL for adverse reproductive effects from this study at 12 mg Pb/kg BW/day. We considered this study as supportive of a NOAEL-based TRV in the range of 1 mg Pb/kg BW/day.

Renal Effects.   The following six additional renal studies support the 1 mg Pb/kg BW/day dose as protective of adverse kidney effects caused by low-level Pb exposure.

Dieter et al. (1993) exposed 6- to 7-week-old male F344/N rats to 0, 30, and 100 mg Pb (as Pb acetate)/kg in food for 30 days. Renal lesions similar to those reported by Fowler et al. (1980) were seen in the 30 and 100 mg Pb treatment groups. A NOAEL of 0.5 mg Pb/kg BW/day was estimated for this study (Table 9.1). However, the data provided by Fowler et al. (1980) suggest that effects at this dose may be of passing significance. The Fowler et al. (1980) study was preferred for TRV development because the study included multigeneration Pb exposures over more sensitive life stages (i.e., fetal, weanling).

Hubermont et al. (1976) exposed adult female Sprague-Dawley rats to 0, 0.1, 1, and 10 mg Pb (as Pb nitrate)/L in drinking water 3 weeks before mating, during pregnancy, and 3 weeks after birth. Pb treatment at 10 mg Pb/L caused a decrease in ALAD activity in blood and kidney, and increased free-tissue porphyrins in kidney of newborns (21 d old). The biological and ecological significance of these endpoints could not be established, and therefore the study was inappropriate for developing a TRV; however, the renal effects occurred near drinking water concentrations reported in Fowler et al. (1980) to alter renal ultrastructure.

Gupta et al. (1995) exposed two-month-old female mice (LACA strain) to daily doses of 0, 6.0, 14, and 28 mg Pb per kg BW by gavage through pregnancy. The highest doses of Pb (14 and 28 mg Pb [as Pb acetate]/kg BW/day) significantly reduced the trace mineral content (i.e., iron, copper, manganese), reduced ALAD activity, and increased glutathione levels in the kidneys. As stated for the Hubermont et al. (1976) study, the biological and ecological significance of these endpoints could not be established. Therefore, we considered this study as inappropriate for developing a TRV; however, the renal effects described occurred within dosing levels previously reported to alter renal ultrastructure.

Khalil-Manesh et al. (1993) exposed 2-month-old Sprague Dawley rats to 0 and 55 mg Pb/L in drinking water for up to 12 months. Following exposure, animals were subsampled over 1-, 3-, 6-, 9-, and 12-month intervals. At 12 months, mild focal tubular atrophy, accompanied by interstitial fibrosis, was evident. The authors stated that Pb exposure produced no significant changes in renal function; however, mild histopathologic lesions were evident at 12 months, suggesting incipient damage to the proximal tubules. Because the study was not as long in duration as Fowler et al. (1980), did not include a sensitive life stage (e.g., fetal, weanling), and only evaluated two doses, the finding of no renal functional effect was, in our opinion, inconclusive.

McMurry et al. (1995) exposed adult wild cotton rats (*Sigmodon hispidus*) to 0, 55, and 550 mg Pb (as Pb acetate)/L in drinking water (*ad libitum*) for up to 13 weeks. Uniform alterations in the renal proximal tubular epithelium were observed in the 550 mg Pb/L treatment group. Cells were enlarged, occasionally necrotic, possessed irregular apical borders, and contained nuclear inclusion bodies. Animals in the 55 mg Pb/L treatment group showed similar lesions, but were free of inclusions. The 55 mg Pb/L treatment group was estimated to have received a dose of 11 mg Pb/kg BW/day, which is within the same order-of-magnitude effects level reported by Fowler et al. (1980).

Osteodystrophic Effects. Two studies, Escribano et al. (1997) and Gruber et al. (1997), demonstrated that Pb can affect both bone density and growth. Both studies were designed to explore whether Pb can contribute to bone disorders in humans. Escribano et al. (1997) exposed 50-day-old female rats to 0 and 9.3 mg Pb (as Pb acetate)/kg feed (ad libitum) for 50 days. Pb was shown to affect the development of the axial skeleton (i.e., vertebrae) and produce a histomorphometric decrease in bone mass. No effect on the longitudinal growth of peripheral long bones (i.e., femur) was observed. Because one Pb concentration was tested (i.e., 9.3 mg Pb/kg), and a dose-response relationship was not established, limited inferences can be made concerning the adverse effects of Pb on bones, as well as the potential for adverse ecological effects. Gruber et al. (1997) exposed adult rats to 0, 55, and 2750 mg Pb/L in drinking water (*ad libitum*) for 365 days. A significant dose-dependent decrease in bone density was observed following one year of exposure in the 55 mg Pb/L treatment group. Similar to Escribano et al. (1997), Gruber et al. (1997) reported that histomorphometric analysis of femur revealed significantly elevated osteoid and resorptive trabecular surface features.

Effects-level doses were estimated as 0.9 and 6 mg Pb/kg BW/day for each study, respectively. We concluded that (1) the relationship between the endpoints evaluated and adverse ecological effects was unknown, (2) the 3 or less doses evaluated in each study did not present a robust dose–response relationship, and (3) the results were of limited usefulness because a NOAEL was not established. Hence, neither of these studies was used to develop a bone-specific TRV. Nevertheless, the studies suggested that adverse effects on bone density were potentially significant at the 6 mg Pb/kg BW/day dose estimated from Gruber et al. (1997).

Neurobehavioral Effects.   Four studies, Bushnell and Levin (1983), Grant et al. (1980), Singh and Ashraf (1989), and Singh (1993), provide direct or indirect evidence that Pb causes neurobehavioral deficits in rats (e.g., learning behavior, reflex development), as observed in humans (ATSDR 1999), at low dose levels. Bushnell and Levin (1983) exposed postpartum male rats to 0, 11.4, and 99.7 mg Pb (as Pb acetate)/L in drinking water (*ad libitum*) for 28 days. The 11.4 mg Pb treatment group showed a significant reduction in choice behavior in a complex maze. We concluded that because the maze choice behavior effect was not dose-responsive, only 3 doses were tested, and a NOAEL dose was not determined, the study should only be considered as supporting evidence for establishing a TRV.

Grant et al. (1980) investigated the effects of Pb on rat postnatal physical and behavioral development. Weaned pups were exposed as described in Kimmel et al. (1980) to 0, 0.5, 5, 25, 50, and 250 mg Pb (as Pb acetate)/L drinking water (*ad libitum*) until sacrifice at 6 to 9 months of age. Food and water consumption was unchanged, but significant decreases in body weight were found in offspring exposed to 50 mg Pb/L and greater. The surface righting and air righting reflexes were significantly delayed in rats exposed to 50 mg Pb/L and unaffected at 5 mg Pb/L. Other reflex development landmarks, including auditory startle and visual placing were unaffected by Pb. Pb, at the highest concentration tested, 250 mg/L, had no effect on locomotor development, activity levels, and motor coordination (i.e., rotorod performance). We concluded that the NOAEL of 1 mg Pb/kg BW/day obtained in this study for behavioral effects is supportive of a TRV; however, the potential ecological significance of delayed surface and air righting reflexes is unknown.

At low doses, Pb produces a variety of biochemical changes in the brain and neurons of small mammals. Singh and Ashraf (1989) and Singh (1993) exposed pregnant rats, 5-day-old rats, or 5-week-old rats to Pb acetate (via gavage at 1.0 mg/kg/day, 5 days per week) for 10 or 20 weeks. Singh and Ashraf (1989) reported significant decreases in brain norepinephrine, GABA, and glutamate decarboxylase, and increases in brain glutamate, glutamine/asparagnine, tyrosine, and monoamine oxidase activity in exposed rats. Singh (1993) showed that chronic prenatal Pb exposure delayed the age-dependent decrease in neuron mRNA expression, ADP-ribosylation, and photoaffinity labeling of the mRNA α_1 subunit. Several studies, including Cory-Slechta et al. (1992), Kala and Jadhav (1995a, b), and Widzowski et al. (1994) also showed alterations in brain neurotransmitters (i.e., dopamine, seratonin), dopaminergic receptors, and neuroglial enzymes. While these effects occurred at doses between 1 and 10 mg Pb/kg BW/day, there were no concurrent functional tests to evaluate the significance of the alterations. As such, these studies were considered inappropriate to set a TRV; nevertheless, they do suggest that Pb begins to affect the nervous system at low doses, within the same range as the functional effects (i.e., radial maze choice behavior, surface righting, air righting) described above.

Immune System Effects. Three studies, Luster et al. (1978), Faith et al. (1979), and McMurry et al. (1995), showed that low doses of Pb produced potentially adverse changes in the humoral and cell-mediated immune systems of small mammals. The Luster et al. (1978) and Faith et al. (1979) studies are companion studies to Kimmel et al. (1980) described previously. Luster et al. (1978) showed that 35- to 45-day-old rats, exposed pre- and postnatally to 25 or 50 mg Pb (as Pb acetate)/L drinking water, had enlarged thymus, reduced circulating levels of immunoglobulin G (IgG), and depressed antibody response to sheep red blood cells (SRBCs). Faith et al. (1979) showed that similarly aged and Pb-exposed rats have suppressed lymphocyte responsiveness to mitogen stimulation and reduced delayed hypersensitivity responsiveness. While significant immunological effects were noted in both studies, a clear dose–response relationship and NOAEL were not established.

McMurry et al. (1995) exposed adult wild cotton rats (*Sigmodon hispidus*) to 0, 55, and 550 mg Pb (as Pb acetate)/L in drinking water (*ad libitum*) for up to 13 weeks. Exposure to the 55 mg Pb/L concentration caused a significant decrease in white blood cell, neutrophil, and eosinophil counts and splenocyte yield; however, the same effects were not found at the higher dose (i.e., lack of dose response). At the high drinking water concentration, Pb altered the proliferative responses of splenocytes to mitogenic stimulation and reduced thymus size, similar to effects reported by Faith et al. (1979). However, Pb did not affect the delayed-type hypersensitivity response or the metabolic activity of macrophages. Differences in species sensitivity and/ or differences in exposure regimes (i.e., rats in the Faith et al. [1979] study were exposed for a much longer duration through sensitive life stages) likely accounted for the observed differences in Pb responsiveness.

The biological or ecological significance of the reported immunological effects was not evaluated in each aforementioned study. For example, in no study were concurrent functional tests performed (e.g., disease-challenge tests). As such, we considered these studies to be inappropriate for TRV development. Nevertheless, the studies suggest that Pb is potential immunotoxicant at concentrations above 25 Pb mg/L in drinking water (i.e., doses greater than 5.6 mg Pb/kg BW/day).

Ocular Effects. Following a review of the literature, we identified a study by Fox et al. (1997) that provides direct and indirect evidence that rat visual acuity is adversely affected by Pb. Fox et al. (1997) exposed lactating female Long-Evans rats to 0, 109, and 1090 mg Pb (as Pb acetate)/L in drinking water. Rat pups were exposed to Pb only via mother's breast milk for 3 weeks until weaning. After 90 days, rats, exposed as pups, showed marked retinal degeneration and thinning. Histologic and electron microscopic examination of the retina showed a 22% loss of rod cells and a 30%

loss of bipolar cells, primarily in the inferior retina. In addition, the authors reported a dose-dependent decrease in rod cell sensitivity, rhodopsin content, reduced cGMP phosphodiesterase activity, and cGMP concentration in the retina. Loss of visual acuity, particularly night vision, was considered a significant adverse and ecologically relevant effect. The lowest effective dose estimated in dams that caused adverse effects in pups was 12.6 mg/kg BW/day (Table 1). We considered this study as supportive of a 1 mg Pb/kg BW/day TRV.

Summary of Studies Rejected as Supportive of a TRV
Upon closer examination, many of the studies selected for further review, including Skoczynska et al. (1993), Cory-Slechta et al. (1983, 1985), Cory-Slechta and Pokora (1995), Barratt et al. (1989), Hayashi (1983), Reiter et al. (1975), and Victery et al. (1982), were rejected from further consideration as supporting a TRV. These studies either utilized toxicological or behavioral endpoints with questionable ecological relevance, showed effects that were transitory or reversible, or failed to demonstrate a clear dose-response relationship (see summary statements in Table 9.1).

We concluded that the Al-Hakkak et al. (1988; see Table 9.1) study should not be used to support a TRV for Pb. Al-Hakkak et al. (1988) investigated the effects of ingestion of Pb monoxide alloy on male mouse reproduction. The authors exposed Balb-C albino Swiss mice to 0, 25, or 50 mg Pb monoxide alloy powder/kg diet for 35 days. The 25 mg alloy/kg diet caused an adverse effect on implantation and litter size. Neither body weights nor the Pb content of the alloy were provided in the study. Therefore, it was not possible to estimate the total dose of elemental Pb ingested in the treatment groups. In addition, the alloy used in the study may have contained other metallic constituents that contributed to the observed toxicity. We concluded that the Azar et al. (1973; see Table 9.1) study should not be used to support a TRV for Pb. The Azar et al. (1973) study involved a three-generation, six-litter rat reproductive study with seven different dose levels. The reduction of weanling weight observed in the study has important implications for reproductive potential in the wild, but occurred at concentrations in the feed that generate an estimated dose outside of the 1 to 10 mg/kg BW/day range. At doses estimated to be less than 10 mg/kg BW/day, blood ALAD activity was reduced. At these doses, the adverse effects associated with reduced ALAD in blood are unknown, as well as the ecological relevance of reduced ALAD. For example, the ATSDR (1999) reports that the adverse effects of decreased ALAD (i.e., observed at doses of Pb below 10 mg/kg BW/day), in the absence of detectable effects on hemoglobin levels and erythrocyte function, are of questionable biological significance.

Finally, we concluded that the Krasovskii et al. (1979) study was not appropriate for TRV derivation. Following our review of more recent toxicological literature for Pb, we found that investigators evaluating similar endpoints (as cited in ATSDR, 1999), have not reproduced the Pb dose-response relationships reported by Krasovskii et al. (1979). For example, the Ronis et al. (1998) study is likely one of the better, recent, studies for identifying a NOAEL for the reproductive toxicity of Pb to rats.

Recommendations for Lead TRV

After consideration of the endpoints, dosing information, evaluation of the experimental results, and limitations of the experiments, we propose an individual health and ecologically protective NOAEL TRV of 1 mg/kg BW/ day, based primarily on the kidney toxicity data contained in Fowler et al. (1980) and supported by at least 20 other studies suggesting that a 1 mg/kg BW/day Pb dose would be protective of reproductive, bone, behavioral, immune, embryonic, renal, and ocular effects seen at doses within one order of magnitude higher.

Predicting Potential Ecological Risks of Lead

Because the TRV is based on lead acetate, a highly bioavailable form of lead, we recommend that the exposure assessment be refined so that site-specific lead bioavailability is estimated or measured. Inorganic lead may be found in the environment in various chemical forms (e.g., lead carbonate, lead oxides, lead sulfate, elemental lead). These chemical forms of lead will have different levels of bioavailability based on their water solubility, amount bound to organic/inorganic matter, and particle size. If lead is identified as a chemical of potential ecological concern (e.g., greater than background or ambient conditions), we recommend refining the level of the analysis as appropriate during the ERA (DTSC 1996a):

Proposed Phased Approach for Assessing the Ecological Risks of Lead

Scoping or Problem Formulation Phase

During the Scoping or Problem Formulation Phase, perform an *in vitro* lead bioaccessibility test on representative soil types from the facility, preferably with low, medium, and high concentrations of lead. Procedures for conducting the analysis are available from the Naval Facilities Engineering Command (NAVFAC 2000). While the *in vitro* lead bioaccessibility test is optimized for evaluating lead availability to pigs and humans, the test is conducted under acidic conditions, which are likely present in the stomach of small mammals. For purposes of a conservative screen, we recommend that the test be conducted at a pH concentration of 1.5 (as recommended by NAVFAC 2000).

Phase I Predictive Assessment. During a Phase I Predictive Assessment, calculate hazard quotients based on the estimated solubility of lead provided in the *in vitro* analysis, concentration of lead in surface water (if an applicable exposure route), and the modeled or estimated uptake of lead in plant and/ or animal food items. First, calculate an incidental soil ingestion exposure point concentration ($EPC_{soil\ lead}$) based on the estimated solubility of site-specific lead as follows:

$$EPC_{soil\ lead} = EPC * Solubility\ Lead_{site}$$

where

$EPC_{soil\ lead}$ = Exposure point concentration for lead in incidentally ingested soil as modified by *in vitro* solubility test

EPC = Exposure point concentration (i.e., typically 95th upper confidence interval on the arithmetic mean) calculated from site-specific soil data

$Solubility\ Lead_{site}$ = Estimated solubility of site-specific lead under *in vitro* test conditions (%).

Second, calculate a food ingestion EPC based on the estimated or measured lead content in food (and water, if applicable). The lead content in food can be estimated with soil-to-biota uptake factors for lead from the literature (Sample et al. 1997). Sum the EPCs for soil and food ingestion. Calculate hazard quotients as follows:

$$HQ = EPC_{sum\ lead}/NOAEL\ TRV$$

where

HQ = hazard quotient for lead based on soil *in vitro* solubility test, food ingestion, and the NOAEL TRV

$EPC_{sum\ lead}$ = Exposure point concentration estimated from the incidental ingestion of soil (as modified by the *in vitro* bioaccessibility test) and concentrations of lead in food items.

If the HQ is less than 1 (based on the NOAEL TRV), we recommend that given the level of uncertainty in the risk analysis is acceptable (e.g., site soil sampling data adequately characterizes potential exposure across a hazardous waste site), a decision is made for no further action with respect to the protection of small mammals at the site. If HQ is greater than 1 (based on the NOAEL TRV), we recommend a Phase II Validation Study as described in DTSC Guidance (1996a).

While uncertainty in the Phase I risk assessment is reduced by more accurately estimating the bioavailable or toxic fraction of lead in soil that is incidentally ingested, the assessment still includes elements of uncertainty. First, the assumption that biologically incorporated lead in prey items (e.g., plants, invertebrates) is 100% bioavailable may be an overly conservative assumption. For example, when a predatory bird species (American kestrel, *Falco sparverius*) was fed a diet of chicks containing high levels of biologically incorporated lead, the toxicity of lead was markedly reduced (approximately tenfold) compared to birds administered an equivalent dose of lead in the form of lead acetate in their drinking water (Custer et al. 1984). Second, the

Phase I assessment may underestimate potential lead exposure because the analysis does not include quantification of lead exposure by the inhalation, dermal, or olfactory bulb (Bench et al. 2001) exposure routes. Lead particles in soil can be inhaled or come in direct contact with the skin, but these exposures are expected to be far less significant than by the ingestion exposure route. Hence, the inhalation and dermal exposure routes are often evaluated qualitatively in the uncertainty analysis of the risk assessment (USEPA 2003). Direct absorption of metals via the olfactory bulb uptake is also not considered in the Phase I risk assessment. This exposure route deserves additional consideration and research, especially for burrowing animals (Bench et al. 2001).

Phase II Validation Study. The purpose of the Phase II Validation Study is to field verify exposure and/or toxicity assumptions made during the predictive assessment. For example, lead uptake factors for plant, invertebrate, and other small mammal prey items can be measured in site-specific, field-collected organisms.

Phase III Impact Assessment. The purpose of an impact assessment for lead would be to determine site-specific media concentrations of lead consistent with no adverse lead effects by intensive field and/or laboratory investigations. For example, this may include the field collection of small mammals for organ residue analyses (Shore and Douben 1994), blood lead levels, and/or biomarker analyses (e.g., bioindicators of exposure and effect as determined by biochemical, physiological, or histopathological analysis). Field studies also may include quantitative measurements of species richness and abundance, population levels, or age-class distribution assessment of each species of concern.

Conclusions

We have presented a process by which ecologically protective TRVs for small mammals may be derived from available laboratory toxicity information. Our approach considered reproduction as a critical endpoint for the protection of small mammal populations from the potential deleterious effects of exposure to hazardous wastes. Nevertheless, we recommend consideration of the totality of potential adverse effects to an organism rather than singularly focusing on reproductive endpoints. While low dose endpoints such as enzyme levels were not considered ecologically relevant or applicable to TRV derivation, others (i.e., behavioral, immune, embryonic, renal, ocular, and structural) were considered highly relevant. TRVs should be developed that are protective of the reproductive capacity of an animal, as well as its ability to survive in the wild. While inbred strains of laboratory animals may be relatively sensitive to toxicants, wild animals already living under a host of environmental pressures (e.g., temperature extremes, food availability, predators, and disease), may be as, or more, susceptible to chemical stressors in their habitats.

We have described the use of TRVs in ERA as a means of identifying potential hazards to small mammals inhabiting hazardous waste sites. By selecting a TRV from the laboratory species that is the most toxicologically sensitive to the constituent of concern, and by selecting toxicological endpoints that are protective of reproduction and survival in the wild, we minimize Type II error (e.g., accepting the null hypothesis that the chemical has no effect on a wild animal when it actually does). Conversely, this approach may increase Type I error (e.g., rejecting the null hypothesis when it is actually true) and potentially overstate ecological risk, particularly for species that may be more tolerant of the constituent of concern. However, because it is impractical to test the toxicological sensitivity of every wildlife species potentially inhabiting a hazardous waste site, the regulatory community must assume that protecting the most sensitive laboratory species is necessary, at minimum, to ensure protection of the most sensitive wildlife species.

Finally, ERA is an iterative process; data gaps or uncertainties in the analysis are reduced following each step in the assessment. A significant source of uncertainty in the risk characterization is the quantification of exposure. Lead, as well as other metals or metalloid compounds in the environment, may be less bioavailable at a hazardous waste site than in a laboratory toxicity study. Bioavailability is a critical component of exposure assessment, and must be appropriately addressed in the ecological risk characterization.

References

Agency for Toxic Substances and Disease Registry (ATSDR). (1999) *Toxicological Profile for Lead* (Update). Division of Toxicology/Toxicology Information Services, Atlanta, GA.

Al-Hakkak, Z.S., Zahid, Z.R., Ibrahim, D.K., al-Jumaily, I.S., and Bazzaz, A.A. (1988) Effects of ingestion of lead monoxide alloy on male mouse reproduction, *Archives of Toxicology* 62(1), 97–100.

Azar, A., Trochimowicz, H.J., and Maxfield, M.E. (1973) Review of lead studies in animals carried out at Haskell Laboratory: Two year feeding study and response to hemorrhage study, in *Environmental Health Aspects of Lead: Proceedings, International Symposium*, October 1972, Amsterdam, D. Barth, A. Berlin, R. Engel, et al., Eds., Luxembourg: Commission of the European Communities, Directorate General for Dissemination of Knowledge, Center for Information and Documentation.

Barratt, C.L., Davies, A.G., Bansal, M.R., and Williams, M.E. (1989) The effects of lead on the male rat reproductive system, *Andrologia* 21(2), 161–6.

Bench, G., Carlsen, T.M., Grant, P.G., Woollett, J.S., Jr, Martinelli, R.E., Lewis, J.L., and Divine, K.K. (2001) Olfactory bulb uptake and determination of biotransfer factors in the California ground squirrel 11 *(Spermophilus beecheyi)* exposed to manganese and cadmium in environmental habitats, *Environmental Science and Technology* 35, 270–7.

Bushnell, P.J. and Levin, E.D. (1983) Effects of zinc deficiency on lead toxicity in rats, *Neurobehavioral Toxicology and Teratology* 5(3), 283–8.

Cory-Slechta, D.A. and Pokora, M.J. (1995) Lead-induced changes in muscarinic cholinergic sensitivity, *Neurotoxicology* 16(2), 337–47.

Cory-Slechta, D.A., Pokora, M.J., and Widzowski, D.V. (1992) Postnatal lead exposure induces supersensitivity to the stimulus properties of a D2-D3 agonist, *Brain Research* 598(1–2), 162–72.

Cory-Slechta, D.A., Weiss, B., and Cox, C. (1983) Delayed behavioral toxicity of lead with increasing exposure concentration, *Toxicology and Applied Pharmacology* 71(3), 342–52.

Cory-Slechta, D.A., Weiss, B., and Cox, C. (1985) Performance and exposure indices of rats exposed to low concentrations of lead, *Toxicology and Applied Pharmacology* 78(2), 291–9.

Custer, T.W., Franson, J.C., and Pattee, O.H. (1984) Tissue lead distribution and hematologic effects in American kestrels (*Falco sparverius* L.) fed biologically incorporated lead, *Journal of Wildlife Diseases* 20(1), 39–43.

Dieter, M.P., Matthews, H.B., Jeffcoat, R.A., and Moseman, R.F. (1993) Comparison of lead bioavailability in F344 rats fed lead acetate, lead oxide, lead sulfide, or lead ore concentrate from Skagway, Alaska, *Journal of Toxicology and Environmental Health* 39(1), 79–93.

Department of Toxic Substances Control (DTSC). (1996a) *Guidance for Ecological Risk Assessment at Hazardous Waste Sites and Permitted Facilities. Part A: Overview*, DTSC, Sacramento, CA. Available <http://www.dtsc.ca.gov/ScienceTechnology/eco.html>

Department of Toxic Substances Control (DTSC). (1996b) *Guidance for Ecological Risk Assessment at Hazardous Waste Sites and Permitted Facilities. Part B: Scoping Assessment*, DTSC, Sacramento, CA. Available <http://www.dtsc.ca.gov/ScienceTechnology/eco.html>

Department of Toxic Substances Control (DTSC). (2000) *EcoNOTE4: Use of Navy/U.S. Environmental Protection Agency (USEPA) Region 9 Biological Technical Assistance Group (BTAG) Toxicity Reference Values (TRVs) for Ecological Risk Assessment*, DTSC, Sacramento, CA. Available <http://www.dtsc.ca.gov/ScienceTechnology/eco.html>

Dilts, P.V., Jr and Ahokas R.A. (1979) Effects of dietary lead and zinc on pregnancy, *American Journal of Obstetrics and Gynecology* 135(7), 940–6.

Escribano, A., Revilla, M., Hernandez, E.R., Seco, C., Gonzalez-Riola, J., Villa, L.F., and Rico, H. (1997) Effect of lead on bone development and bone mass: A morphometric, densitometric, and histomorphometric study in growing rats, *Calcified Tissue International* 60(2), 200–3.

Faith, R.E., Luster, M.I., and Kimmel, C.A. (1979) Effect of chronic developmental lead exposure on cell-mediated immune functions, *Clinical and Experimental Immunology* 35(3), 413–20.

Fowler, B.A, Kimmel, C.A., Woods, J.S., McConnell, E.E., and Grant, L.D. (1980) Chronic low-level lead toxicity in the rat. III. An integrated assessment of long-term toxicity with special reference to the kidney, *Toxicology and Applied Pharmacology* 56(1), 59–77.

Fox, D.A., Campbell, M.L., and Blocker, Y.S. (1997) Functional alterations and apoptotic cell death in the retina following developmental or adult lead exposure, *Neurotoxicology* 18(3), 645–64.

Goyer, R.A., Leonard, D.L., Moore, J.F., Rhyne, B., and Krigman, M.R. (1972) Lead dosage and the role of the intranuclear inclusion body. An experimental study, *Archives of Environmental Health* 20(6), 705–11.

Grant, L.D., Kimmel, C.A., West, G.L., Martinez-Vargas, C.M., and Howard, J.L. (1980) Chronic low-level lead toxicity in the rat. II. Effects on postnatal physical and behavioral development, *Toxicology and Applied Pharmacology* 56(1), 42–58.

Gruber, H.E., Gonick, H.C., Khalil-Manesh, F., Sanchez,T.V., Motsinger, S., Meyer, M., and Sharp, C.F. (1997) Osteopenia induced by long-term, low- and high-level exposure of the adult rat to lea, *Mineral and Electrolyte Metabolism* 23(2), 65–73.

Gupta, G.S., Singh, J., and Parkash, P. (1995) Renal toxicity after oral administration of lead acetate during pre- and post-implantation periods: Effects on trace metal composition, metallo-enzymes and glutathione, *Pharmacology and Toxicology* 76(3), 206–11.

Hayashi, M. (1983) Lead toxicity in the pregnant rat. II. Effects of low-level lead on delta-aminolevulinic acid dehydratase activity in maternal and fetal blood or tissue, *Industrial Health* 21(3), 127–35.

Hubermont, G., Buchet, J., Roels, H., and Lauwerys, R. (1976) Effect of short-term administration of lead to pregnant rats, *Toxicology* 5(3), 379–84.

Junaid, M., Chowdhuri, D.K., Narayan, R., Shanker, R., and Saxena, D.K. (1997) Lead-induced changes in ovarian follicular development and maturation in mice, *Journal of Toxicology and Environmental Health* 50(1), 31–40.

Kala, S.V. and Jadhav, A.L. (1995a) Low level lead exposure decreases in vivo release of dopamine in the rat nucleus accumbens: A microdialysis study, *Journal of Neurochemistry* 65(4), 1631–5.

Kala, S.V. and Jadhav, A.L. (1995b) Region-specific alterations in dopamine and serotonin metabolism in brains of rats exposed to low levels of lead, *Neurotoxicology* 16(2), 297–308.

Khalil-Manesh, F., Gonick, H.C., and Cohen, A.H. (1993) Experimental model of lead nephropathy, continuous low-level lead administration, *Archives of Environmental Health* 48(4), 271–8.

Kimmel, C.A., Grant, L.D., Sloan, C.S., and Gladen, B.C. (1980) Chronic low-level lead toxicity in the rat. I. Maternal toxicity and perinatal effects, *Toxicology and Applied Pharmacology* 56(1), 28–41.

Krasovskii, G.N., Vasukovich, L.Y., and Chariev, O.G. (1979) Experimental study of biological effects of lead and aluminum following oral administration, *Environmental Health Perspectives* 30, 47–51.

Luster, M.I., Faith, R.E., and Kimmel, C.A. (1978) Depression of humoral immunity in rats following chronic developmental lead exposure, *Journal of Environmental Pathology and Toxicology* 1(4), 397–402.

McMurry, S.T., Lochmiller, R.L, Chandra, S.A., and Qualls, C.W., Jr. (1995) Sensitivity of selected immunological, hematological, and reproductive parameters in the cotton rat (*Sigmodon hispidus*) to subchronic lead exposure, *Journal of Wildlife Diseases* 31(2), 193–204.

Naval Facilities Engineering Command (NAVFAC). (2000) *Guide for Incorporating Bioavailability Adjustments into Human Health and Ecological Risk Assessments at U.S. Navy and Marine Corps Facilities. Part 2: Technical Background Document for Assessing Metals Bioavailability*, NFESC User's Guide, UG-2041-ENV, NAVFAC, Washington, DC. Available <http://erb.nfesc.navy.mil/erb_a/support/wrk_grp/bio_a/bioa_guide_final2.pdf>

Reiter, L.W., Anderson, G.E., Laskey, J.W., and Cahill, D.F. (1975) Developmental and behavioral changes in the rat during chronic exposure to lead, *Environmental Health Perspectives* 12, 119–23.

Ronis, M.J., Badger, T.M., Shema, S.J., Roberson, P.K., and Shaikh, F. (1998) Effects on pubertal growth and reproduction in rats exposed to lead perinatally or continuously throughout development. *Journal of Toxicology and Environmental Health Part A* 53(4), 327–41.

Sample, B.E., Aplin, M.S., Efroymson, R.A., Suter, G.W., II, and Welsh, C.J.E. (1997) *Methods and Tools for Estimation of the Exposure of Terrestrial Wildlife to Contaminants*, Oak Ridge: Oak Ridge National Laboratory, Oak Ridge, TN. Available <http://www.esd.ornl.gov/programs/ecorisk/documents/tm13391.pdf>

Shore, R.F. and Douben, P.E.T. (1994) Predicting ecotoxicological impacts of environmental contaminants on terrestrial small mammals, in *Reviews of Environmental Contamination and Toxicology, Volume 134*, G. Ware, Ed., New York: Springer-Verlag New York, Inc.

Sierra, E.M. and Tiffany-Castiglioni, E. (1992) Effects of low-level lead exposure on hypothalamic hormones and serum progesterone levels in pregnant guinea pigs, *Toxicology* 72(1), 89–97.

Singh, A.K. (1993) Effects of chronic low-level lead exposure on mRNA expression, ADP-ribosylation and photoaffinity labeling with [alpha-32P]guanine triphosphate-gamma-azidoanilide of GTP-binding proteins in neurons isolated from the brain of neonatal and adult rats, *Biochemical Pharmacology* 45(5), 1107–14.

Singh, A.K. and Ashraf, M. (1989) Neurotoxicity in rats sub-chronically exposed to low levels of lead, *Ecotoxicology and Environmental Safety* 31(1), 21–5.

Skoczynska, A., Smolik, R., and Jelen, M. (1993) Lipid abnormalities in rats given small doses of lead, *Archives of Toxicology* 67(3), 200–4.

United States Environmental Protection Agency (USEPA). (1994) *Guidance for the Data Quality Objectives Process*, Quality Assurance Management Staff, USEPA, Washington, DC. Available <http://www.epa.gov/quality/qs-docs/g4-final. pdf>

United States Environmental Protection Agency (USEPA). (1997) *Ecological Risk Assessment Guidance for Superfund: Process for Designing and Conducting Ecological Risk Assessments*, Office of Solid Waste and Emergency Response, USEPA, Washington, DC. Available <http://www.epa.gov/superfund/programs/risk/ecorisk/ecorisk.htm>

United States Environmental Protection Agency (USEPA). (1998) *Guidelines for Ecological Risk Assessment*, USEPA Risk Assessment Forum, Washington, DC. Available <http://cfpub.epa.gov/ncea/cfm/recordisplay.cfm?deid=12460>

United States Environmental Protection Agency (USEPA). (2003) *Guidance for Developing Ecological Soil Screening Levels*, Office of Solid Waste and Emergency Response, USEPA, Washington, DC. Available <http://www.epa.gov/superfund/programs/risk/ecorisk/ecossl.htm>

Victery, W., Vander, A.J., Markel, H., Katzman, L., Shulak, J.M., and Germain, C. (1982) Lead exposure, begun *in utero*, decreases renin and angiotensin II in adult rats, *Proceedings of the Society for Experimental Biology and Medicine* 170(1), 63–7.

Widzowski, D.V., Finkelstein, J.N., Pokora, M.J., and Cory-Slechta, D.A. (1994) Time course of postnatal lead-induced changes in dopamine receptors and their relationship to changes in dopamine sensitivity, *Neurotoxicology* 15(4), 853–65.

Wiebe, J.P. and Barr, K.J. (1988) Effect of prenatal and neonatal exposure to lead on the affinity and number of estradiol receptors in the uterus, *Journal of Toxicology and Environmental Health* 24(4), 451–60.

Index

A

Agency for Toxic Substances and Disease
  Registry (ATSDR), 6
aluminum, *see also* aluminum chloride, *see also*
    aluminum hydroxide, *see also*
    aluminum nitrate nonahydrate, *see also*
    aluminum sulfate, 66–72. *see also*
    aluminum sulfate
  aggression in animal studies and, 68
  avoiding high-dose consumption during
    gestation and lactation, 72
  bone retardation in animal studies and, 71
  chelating agents and, 71
  developmental toxicity and, 70, 71
  developmental toxicity in animal studies
    and, 71
  effect on steroidogenesis, 68
  embryo/fetal toxicity and, 6, 68, 68–69, 70
  exposure during pregnancy as
    developmental hazard, 72
  exposure during pregnancy in humans, 70
  fertility in animal studies and, 68
  gastrointestinal absorption and diet, 70
  high-dose antacids during pregnancy and,
    67
  intrauterine development in mammals
    and, 65–92
  maternal stress and morphological defects,
    71
  maternal stress in animal studies and, 70,
    71
  maternal toxicity in animal studies and, 71
  nitric oxide products and, 68
  NOAEL (no observable adverse effect
    level, 67
  reduced gastrointestinal absorption with
    oral silicon, 72
  reproductive toxicity and, , 67, 67–72, *69t*

  reproductive toxicity in animal studies, 67
  sexual behavior in animal studies and, 68
aluminum chloride. *see also* aluminum
  embryo/fetal toxicity in animal studies
    and, 68
  teratogenic effects in animal studies and,
    68
aluminum hydroxide. *see also* aluminum
  developmental toxicity and diet, 70
  poor absorption and, 68
  skeletal variations and, 70
aluminum nitrate
  embryo/fetal toxicity and, 70
  malformations in animal studies and. *see*
    also Al nitrate nonahydrate
aluminum nitrate nonahydrate. *see also*
    aluminum
aluminum sulfate. *see also* aluminum
  water supply and reproductive toxicity
    and, 67
American Conference of Industry Hygienists,
  202
aquatic organisms, 25–28
  organotin compounds and, 25–28
arsenate, *see also* arsenic, 13
arsenic, *see also* arsenate, *see also* arsenite, *see*
    also trace metals, 12–17, 13. *see also*
    arsenite
  animal studies of effects on steroid
    hormone production, 6
  carcinogenesis in animal studies, 168
  cell death and, 13
  concentrations in drinking water and, 14
  detoxification via methylation and, 13
  developmental toxicity and, 17
  disruption of PAX3 gene with neural tube
    closure and, 13
  environmental contamination and, 13
  estrogenic activity and, 17

245